Traffic Radar Handbook

A Comprehensive Guide to Speed Measuring Systems

includes microwave and laser police traffic radar

By

Donald S. Sawicki
CopRadar.com

© 2002 by Donald S. Sawicki. All rights reserved.

No part of this book may be reproduced, stored in a retrieval system, or transmitted by any means, electronic, mechanical, photocopying, recording, or otherwise, without written permission from the author.

ISBN: 0-7596-8832-X

This book is printed on acid free paper.

Table of Contents

Preface V

Chapter 1 — RADAR/SPEED MEASURING SYSTEMS
1.1 — Police Traffic Radars 1
1.2 — Other Speed Measuring Systems 8

Chapter 2 — THEORY
2.1 — The Doppler Principle 12
2.2 — Radar Cross Section (RCS) 18

Chapter 3 — RADAR HARDWARE 29
3.1 — The Basic Radar Gun 29

Chapter 4 — THE COSINE EFFECT (ERROR)
4.1 — Cosine Error Geometry 39
4.2 — Cosine Effect On Measured Speed 43
4.3 — Cosine Effect On Moving Radar 49
4.4 — Cosine Effect On Across The Road Radar 54
4.5 — Targets On A Curve 56

Chapter 5 — LIMITATIONS AND PROBLEMS WITH RADAR
5.1 — NHTSA 1980 Tests 60
5.2 — Test / Calibration Before Use 64
5.3 — Beam / Antenna Limitations 69
5.4 — Operational Problems 76
5.5 — Interference 86

Chapter 6 — LASER RADAR (LADAR OR LIDAR)
6.1 — Introduction To Laser Radar 95
6.2 — Light Spectrum 97
6.3 — Laser Radar Operation 99
6.4 — Laser Radar Aiming / Alignment 105
6.5 — Operational Problems With Laser Radar 108
6.6 — Laser Radar Speed Error (due to range error) 111

Chapter 7 — BEFORE/AFTER AN ENCOUNTER WITH RADAR/LADAR
7.1 — Countermeasures 118
7.2 — The Courtroom 130

Chapter 8 — **RADIO FREQUENCY (RF) TOPICS**
8.1 — RF Biological Effects 141
8.2 — RF Radiation Standards 147
8.3 — Traffic Radar Power Density 150

Appendix A — **Frequency Spectrum** 153
Appendix B — **Electromagnetic Waves** 159
Appendix C — **Doppler Equations** 163
Appendix D — **Radar Range Equation** 167
Appendix E — **Radar Technical Details** 169
Appendix F — **Ladar Technical Details** 175

Constants / Conversions / Abbreviations 179
Additional Reading / Sources 183

Preface

The mere subject of police radar is as controversial today as it was in the fall of 1973 when the 55 mph national speed limit started (a temporary law that lasted 22 years) and solid-state traffic radars started showing up on the roads in large numbers. The advantage of solid-state (versus vacuum tubes) was to bring down the cost, size, and power requirements of the units. Cost reductions included not having a strip chart recorder for a written record of target history. Even through the temporary 55 mph national speed limit is history (1973 - 1995 NOV), thanks to Republican control (really Democrat loss of control) of both the U.S. Senate and House, many communities continue to use and abuse traffic radar because these communities have become accustomed to the revenue obtained from the use of radar. Thus traffic radar continues to be used extensively across the nation.

Many police departments and agencies have greatly improved their application of radar to traffic control, but there still exists a large number not qualified to operate this type of instrument. To operate a traffic gun does not require genius, but it does require proper training as well as a basic understanding of the device. Hopefully the reader will gain enough understanding of traffic radar to help prevent or right any injustices these instruments (really the police, courts, elected or anointed bureaucrats) may impose.

Vehicle Insurance Companies also have an interest in traffic radar. Insurance companies base automobile insurance rates on driving record; the more speeding tickets a driver has the higher the insurance rate. In other words, the more speeding tickets issued the more money the insurance companies make.

Traffic Radar Handbook
A Comprehensive Guide to Speed Measuring Systems

Chapter 1.1

Police Traffic Radars

Outline
- General Description
- S Band (obsolete)
- X Band
- K Band
- Ka Band
— Wideband (Ka) Radar
- Across the Road Photo / Safety Radar
- Laser Radar
- Microwave versus Laser Radar

General Description
    radar (ra'där), noun—(1) acronym for **RA**dio Detection And Ranging. (2) a remote sensor that emits electromagnetic waves (radio, microwave, or light) in order to measure reflections for detection purposes (presence, location, motion, etc.). (3) radiolocation. (4) field disturbance sensor.
    All stationary microwave traffic radars measure on-coming traffic; some models also (and/or) measure going (receding) traffic. Virtually all moving mode radars can operate from a stationary position or a moving patrol car. Moving mode radars usually require a minimum patrol car speed for moving mode operation. All moving mode radars (in moving mode) measure on-coming (opposite lane) traffic; some can also (and/or) measure receding (opposite lane) traffic (requires aft antenna). Some moving mode radars can measure targets traveling in the same-lane (direction) as the patrol car (front and/or rear antenna). Same-lane radars require a minimum speed difference (2 mph or MORE) between the target and patrol car. Moving radars also measure patrol car speed (ground speed), most also display patrol measured speed. Many radars track only **one target** at a time; some models have the option to track and display **two targets**—the strongest (may be closest or largest) target (echo) and the fastest (or a faster) target in the beam.
    Most microwave traffic radars have a relatively wide **beam** (10 to 25 degrees) that easily covers several lanes of traffic at a relatively short range (see chapter 5.3—Beam /Antenna Limitations). Detection range (in the beam) varies with radar and target reflectivity and may be as low as 100 feet (30 meters) or

1

less to 1 mile (1.6 kilometers) or more. A radar may track a distant large target instead of a closer small target without any indication to the operator which target the radar is tracking (see chapter 2.2—Radar Cross Section).

The angle between the (microwave or laser) radar and target (**alpha** in the figure below) must be small for the radar to accurately measure speed. The angle is referred to as the **Cosine Effect** angle because measured speed is directly proportional to the cosine of this angle; the larger the angle the lower the measured speed (see chapter 4.1—Cosine Effect Geometry). The radar should be located as close to the road (really projected target path) as practical to minimize Cosine Effect errors (see chapter 7.2—The Courtroom / Cosine Effect Defense). The Cosine Effect on moving mode radar is slightly more complicated and may measure target speed high under certain conditions (see chapter 4.3—Cosine Effect on Moving Radar).

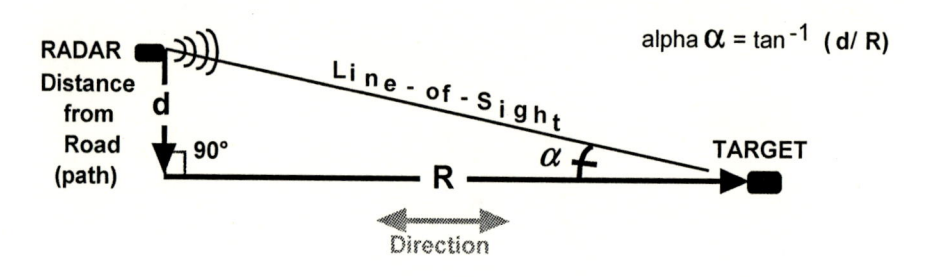

RADAR Distance from Road (path) — d — 90° — Line-of-Sight — R — α — TARGET

alpha $\alpha = \tan^{-1}(d/R)$

Direction

Radar Measured Target Speed = Target Speed x *cos* α

Figure 1.1-1—**Radar Set-up**

Some microwave radars constantly transmit. Some microwave and all laser traffic radars only transmit on operator command (**instant-on**); some microwave radars (**pulsed**) only transmit periodically every several seconds, and then only long enough (fractions to 1 second or so) to get a speed measurement. Instant-on and pulsed (microwave) radars are intended to defeat radar detectors by keeping transmission time short.

Some microwave radars and many laser radars have a timing mode that allows the operator to **time targets** (traveling between 2 points of known distance) visually instead of transmitting (defeats microwave/laser detectors). This method takes more time to set-up, requires more operator actions, is less versatile, and thus used less often. See chapter 1.2—Other Speed Measuring Systems Timing Computer.

Also see;
chapter 3.1—Basic Radar Gun
chapter 6.3—Laser Radar Operation

S Band Radar (obsolete)

One of the first traffic radars was built in 1947 by a Connecticut firm (Automatic Signal Co.) and used by Connecticut State Police on Route 2 in Glastonbury. Early radars were bulky and heavy systems (vacuum-tube technology) that usually consisted of three or more separate pieces of equipment, an antenna (sometimes 2 antennas—separate transmit and receive), a 45 pound (20 kg) box (the tube transmitter, receiver and processor), a strip chart pen recorder for a permanent record, and a needle meter calibrated in mph. Sometimes the antennas mounted on a tripod and sometimes on the hood or fender of a patrol car. Some of the early 1960s' models mounted the antennas in the back windshield of the patrol car.

The first traffic radars transmitted at **2.455 GHz** in the S band (2 - 4 GHz); note that many microwave ovens transmit at about 2.45 GHz. Radar antenna beamwidths varied from 15 to 20 degrees depending on model. These radars operated from a stationary position only and measured receding as well as approaching targets to an accuracy of about ± 2 mph. The maximum detection range was an unimpressive 150 to 500 feet (45 to 150 meters); vacuum-tube receivers do not have the sensitivity of solid-state receivers. A radar with a 150 foot detection range would have less than 1.5 seconds to measure a target traveling 68 mph (100 feet/second or 109 kmh). S band radars are obsolete.

X Band Radar

Frequency ± Tolerance - - Frequency Range - -
10.525 GHz ± 25 MHz - - 10.500 - 10.550 GHz - -

X band radars have been around since 1965 and operate on a single frequency (one 50 MHz channel). Radars in the X band have better all weather performance (less signal attenuation in bad weather) than K or Ka bands. X band radars tend to have wider beams than K or Ka radars. Models that measure targets faster then 420 mph (or 420 kmh) may sacrifice accuracy (± 1 mph or kmh), unless the frequency tolerance (± 25 MHz) is tighten.

Donald S. Sawicki

K Band Radar

Frequency ± Tolerance - - Frequency Range - -
24.150 GHz ± 100 MHz - - 24.050 - 24.250 GHz - -
or
24.125 GHz ± 100 MHz - - 24.025 - 24.225 GHz - -

K band radars have been around since 1976 and operate on a single frequency (one 200 MHz channel). Models that measure targets faster then 240 mph (or 240 kmh) may sacrifice accuracy (± 1 mph or kmh), unless frequency tolerance (± 100 MHz) is tighten.

Side note: Some World War II radars operated in the K band around 24.1 GHz (in the limits of K band traffic radar), which also happens to be in the water vapor absorption band (centered at about 22.24 GHz); signals in the absorption band tend to become absorbed by moisture in the atmosphere and do not have the range that other frequency bands offer. For short range applications the effects may be tolerable on relatively clear dry days.

Ka Band Radar

**Traffic Radar Ka Band
33.4 - 36.0 GHz**

In 1983 the U.S. FCC allocated the spectrum from 34.2 - 35.2 GHz (Ka band) for traffic radar use, that same year Ka band photo (Across the Road) radars started appearing in the United States. In 1992 the FCC expanded the Ka band spectrum allocated for traffic radar use to 33.4 - 36 GHz.

Target detection range dependents on moisture in the atmosphere (rain or humidity), the more moisture the less range. Models that measure targets faster then 333 mph (or 333 kmh) for a 33.4 GHz radar or 359 mph (or 359 kmh) for a 36 GHz radar, may sacrifice accuracy (± 1 mph or kmh), unless frequency tolerance (± 100 MHz) is tighten.

The available bandwidth allocated to Ka band traffic radar is 2.6 GHz (36-33.4 GHz), which equals 2,600 MHz. Most Ka traffic radars have a frequency tolerance of ± 100 MHz (200 MHz bandwidth); each channel allocated 200 MHz of bandwidth. Therefore 2,600 MHz (Ka bandwidth available) divided by 200 MHz (channel bandwidth) equals 13 (channels). A traffic radar in the Ka band with a frequency tolerance of ± 100 MHz may have more channels, but some or all the channels will overlap. Some models transmit on a single frequency only;

Traffic Radar Handbook
A Comprehensive Guide to Speed Measuring Systems

other models allow the operator to select one of several fixed frequencies (see appendix A, table A-3—Select Ka Band Traffic Radar Frequencies).

Wideband (Ka) Radar

Wideband radars (Ka band) operate on a single fixed frequency (operator selects one of several available), and/or in a frequency hop mode. In frequency hop mode the radar dwells on a fixed frequency for a fraction of a second (on the order of 1/10 of a second or more) and hops to another frequency. The radar cycles between a number of different frequencies. Wideband radars are intended to defeat radar detectors (see appendix A, figure A-1—Select Ka Band Radars).

Across the Road (Photo / Safety) Radar

ACROSS THE ROAD (photo or safety) radars are designed to point a narrow beam (typically 5 degrees) across the road at an angle—instead of down the road. The beam cannot cross the road at anything close to 90 degrees but something much less (typically about 22 degrees). The main beam of the radar paints only a small portion of the road. These systems, if designed properly, account for the Cosine Effect angle (based on alignment angle and beamwidth) and adjust (upward 6% to 9% for a 5° beam aligned at 22°) measured target speed.

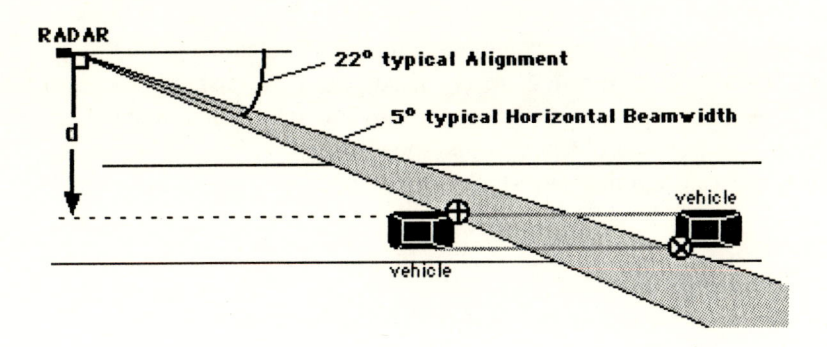

Figure 1.1-2—**Across the Road Radar Set-up**

Photo radars, or camera radars as they were first called, were in experimental stages of development as early as 1954 using S band radars. In 1983 the state of Texas tried a French-manufactured Ka band radar for a time but discontinued its use because the units were being stolen right off the road. Many communities continue to use photo radar because of the revenue it generates;

some communities have outlawed photo radar, usually because of public pressure.

Photo radar (constantly transmitting) automatically detects a speeding violation (auto-lock) and photographs and/or video tapes the suspect vehicle, and records vehicle speed and typically the date, time and location. In states that have only one license plate (rear) two photographs (and cameras) are required to photograph the license plate and driver—one photo (front of vehicle) to get the driver, another photo (rear of vehicle) to capture the license plate. Some units function at night by using a flash. Some units use an orange flash filter—an orange flash is not as bright (as a white light flash) and should startle the driver less. Many photo radars connect to a computer or printer to retrieve stored data statistics such as number of violations, time and speed of each violation, etc.

A police officer does not have to see (or even be near) the alleged violation—the process is automatic. The police officer is replaced by electronic circuits and a still and/or video camera. Many drivers do not even know they were recorded by photo radar (usually hidden in a van, pick-up truck, highway maintenance vehicle, etc.) until weeks later when a ticket and photograph (usually includes license plate and driver) arrives in the mail. Note that the registered vehicle owner may not be the driver, but the owner still gets the ticket.

Some Across the Road Radars connect to a large display to indicate a driver's speed to the driver (Safety Radar). Some Safety Radars only display speeds above the limit, some display all speeds measured. Some units record all speeds measured, some only record speed violations. Safety Radars usually do not record the driver or vehicle, unless connected to a camera.

When operator controlled, rarely used, the radar can be set to constantly transmit, or set to instant-on (transmit only on operator command). In either mode the operator can observe target speed on a display and/or set the radar to automatically photograph (auto-lock) violations.

Also see;
chapter 4.1—Cosine Error Geometry,
chapter 4.4—Cosine Effect on Across the Road Radar.

Laser Radar

Laser radars (Chapter 6), sometimes referred to as ladars (**LA**ser Detection **A**nd **R**anging) or lidars (**LI**ght **D**etection **A**nd **R**anging), were introduced in the early 1990s. These systems radiate in the upper infrared (IR) band and have extremely narrow beams (compared to microwaves). Laser radars function from a stationary position only (no moving mode) and measure approaching and/or receding targets, some also display target range. Most laser radars can also measure the range of stationary objects.

Microwave versus Laser Radar

Microwave traffic radars do not require the operator to aim exactly (only general direction) at one particular target, and are most effective when traffic is light. Many microwave radars can be used from a moving patrol car. Traffic laser radars can function in dense (or light) traffic, and require the operator to select (aim crosshairs at) a particular target. Laser radars are not designed to operate from a moving patrol car. Typically microwave radars have longer detection range than laser light systems.

MICROWAVE RADAR	**LASER RADAR**
Stationary or Moving*	Stationary
Easy Aim	Exact Aim Required
Continuous, Instant-on*, Intermittent (Pulsed)*	Instant-on only
Light - Moderate Traffic	Light - Dense Traffic
Short - Long Range Targets	Short Range Targets
Measures Speed Only some models measure / display Strongest and/or Fastest* targets	Measures Speed and Range some models do not display Range
Inside or Outside Patrol Car antennas can be mounted / used inside or outside* patrol car (see NOTE below)	Outside Patrol Car Use Should not be operated from behind glass / windshields etc.

* optional mode / feature

NOTE: Do Not Use a radar, or a radar detector, from behind an **electrically heated windshield** such as the Ford Instaclear or General Motors (GM) Electriclear—these windshields have metal film coatings that block microwave signals. Also note vehicles with these (reflective) windshields have a larger radar cross section—increasing detection range for a radar.

Donald S. Sawicki

Chapter 1.2

Other Speed Measuring Systems

Outline
- Across the Road Laser
- Aerial Clocking
- Pacing (speedometer)
- Road Cables
- Inductive Loops
- Stop Watch
- Timing Computer
- VASCAR

Except for pacing and microwave radar (which uses Doppler—chapter 2.1), most speed measuring systems use a form of distance/time (clocking) to compute speed. Distance = Speed x Time

Speed = Distance / Time

Figure 1.2-1—
Across the Road Laser

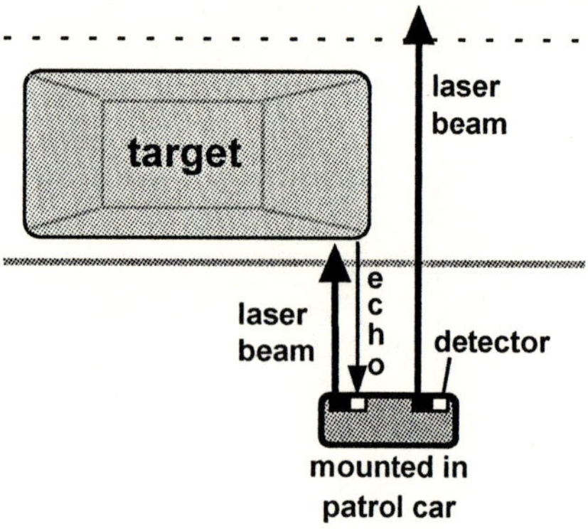

Across the Road Laser
Across the road laser uses two independently pulsed (prevents mutual interference) laser beams that cross the road at a 90 degree angle. The system measures the time it takes for the front of a vehicle to travel from the first beam to the second beam; speed equals distance (between beams) divided by time (for target to travel from first to second beam). The time it takes the target to clear (rear of target out of beam) the first and second beam is used to compute a second speed measurement. The two speed readings must match closely (or the data is rejected) in the event multiple targets pass through the beams at the same time.

Sodi Scientifica SpA, of Florence, Italy builds the Autovelox 104-C, an across the road laser that uses two laser beams spaced 0.4 meters (15.7 inches) apart. The system has an optional 35 mm camera for automatic operation.

Aerial Clocking

Aerial Clocking is probability the best system for police in terms of catching large numbers of speeders, but one of the worst in terms of personnel and equipment (expense). A traffic officer observes traffic from an airplane or helicopter and times targets as they pass between two points of known distance—usually a series of lines painted across the road spaced 1/10 mile (0.1 miles = 176 yards = 528 feet) apart. The airborne officer times targets with a stop-watch and time-speed-distance charts (or a calculator), or a Timing Computer (see below). The officer radios to patrols on the ground which vehicle to ticket when a violation is observed.

Pacing (speedometer)

Pacing involves following a target vehicle at a constant range; the longer the officer follows a target, the better the chance of an accurate measure of target speed. This method depends on the ability of the officer to maintain patrol car speed (constant distance between patrol car and target) with target speed, and the accuracy of the patrol car speedometer. The speedometer works by measuring drive shaft rotation either by digital or analog means; accuracy is good to about ± 1 mph. Accuracy may be compromised if tire pressure changes or different size tires or wheels are used, or the rear axle load is different from installed specifications. NOTE, all speedometers have maximum limits (some as low as 70 mph) and some have minimum limits (as high as 20 mph). Some police agencies recommend only pacing forward vehicles (not vehicles in rear view mirror), and tracking (constant separation) target for at least 0.2 miles (over 320 meters) to get an accurate speed reading (± 2 mph).

Road Cables

Road cable systems are less popular because of expense, relatively large size, and long setup time. Two or more pneumatic cables are placed a fixed distance apart across the road; target speed is determined by measuring the time it takes a target to travel between the cables.

ESP (Excessive Speed Preventor) is one road cable system; ORBIS is another (earlier) system that also automatically photographs detected violations. Early (1996-97) versions of TruVel (UK) used 3 cables; later versions used 4 cables.

Donald S. Sawicki

Inductive Loops

Inductive loop systems use 2 sensors (buried in road or laid rubber strips) spaced a fixed distanced apart. An inductive loop senses a vehicle by detecting changes in magnetic field induced by a vehicle on/above the sensor (inductive loops are also used to trigger some stoplights). Target speed is determined by measuring the time it takes a target to travel between the 2 sensors. TruVelo Combi S (UK) is an inductive loop system that interfaces with a photo / IR flash for automatic violation detection.

Stop Watch

A stop watch can be used to calculate speed by measuring the time it takes a target to travel between 2 points (of known distance). The greater the distance the more accurate the measurement. Any stop-watch used to time targets should be calibrated by a qualified laboratory.

To compute speed in **mph** from distance (**feet**), and time (**seconds**);
Speed = (**15/22**) x Distance / TIME

To compute speed in **kmh** from distance (**meters**), and time (**seconds**);
Speed = **3.6** x Distance / TIME

Distance / Timing Computer

Hand held Distance / Time computers (such as the Robic SC-514 Stopwatch—UK), some microwave radars, and many laser radars incorporate a stop-watch or timing mode. The operator inputs distance (typically feet or meters) between points and uses the computer to time targets (by pushing a button when target passes each point). Target speed calculated is based on time (between button pushes) and distance (previously entered). A radar or ladar in Distance / Time mode does not transmit microwave or light radiation (defeats radar / ladar detectors). Most laser radars have a range only mode (to measure distance to an object) to assist determining distance between points (distance entered into ladar timing computer).

VASCAR

VASCAR (Visual Average Speed Computer And Recorder) uses operator inputs to a computer to determine target speed. This requires multiple operator inputs or some pre-determined information. Basically, VASCAR uses the **distance** between two arbitrary points and the **time** a target takes to travel between these two points to determine target speed (a fancy stop-watch or timing computer). The operator inputs the time parameters by pushing a button when the target passes each point. If the distance is not already known the operator drives by each point to measure the distance; a button is pushed when the driver passes each point so VASCAR can compute distance using the patrol car odometer cable.

VASCAR plus
—Displays speed readings 0 to 120 mph (± 1.0 percent)
—Operator can display time or distance (miles or feet)
—Input distance up to 0.999 miles
—Measures distances from 1 foot to 100 miles
—Time measured in increments of 0.036 seconds up to 10 hours
—metric versions available
—memory clip contains system software
DC power: 10 to 18 Vdc (1 A max)
Operating temperatures: -20 to 120 degrees Fahrenheit
Control Box: 6 3/4 x 3 3/8 x 4 1/8 inches
Odometer module 1 1/4 x 1 x 3 1/2 inches
Manufactured by: Traffic Safety Systems Inc, 3206 Lanvale Ave, PO Box 27306, Richmond, VA 23261, 804-353-3826

FEDERAL VASCAR
—Displays speed readings up to 199.9 mph
—Displays speed to tenths of a mile
—Measures distance to a nominal 5.46 miles
—Measures time to a nominal 6.39 minutes
Manufactured by: Federal Sign and Signal Corporation, Signal Division, 136th and Western Ave., Blue Island, IL 60406, 312-468-4500 Chicago, 312-389-3400 Blue Isla

Donald S. Sawicki

Chapter 2.1

The Doppler Principle

Outline
- Doppler Examples
- Stationary Radar Doppler
- Moving Radar Doppler

Doppler Examples
Everyday life has multiple examples of the Doppler phenomenon with sound; the whistle from a moving train is a good example. As the train approaches a stationary listener, the pitch (frequency) of the whistle sounds higher than when the train passes by, at which time the pitch sounds the same as if the train were stationary. As the train recedes from the listener, the pitch decreases. Car horns exhibit the same phenomenon, as does all sound. Note that in the above example if a car horn is stationary and a listener is on the train, the Doppler principle still applies. As the listener on the train approaches the stationary horn, the pitch of the horn sounds higher; as the train recedes from the stationary horn the pitch sounds lower (to anyone on the train).

Electromagnetic waves radiated by traffic radar, as well as sound waves, obey the Doppler principal, although electromagnetic waves travel at the speed of light and audio waves travel at the speed of sound. The Doppler effect is a frequency shift that results from relative motion between a frequency source and a listener. If both source and listener are not moving with respect to each other (although both may be moving at the same speed in the same direction), no Doppler shift will take place. If the source and listener are moving closer to each other, the listener will perceive a higher frequency—the faster the source or receiver is approaching the higher the Doppler shift. If the source and listener are getting farther apart, the listener will perceive a lower frequency—the faster the source or receiver is moving away the lower the frequency. The Doppler shift is directly proportional to speed between source and listener, frequency of the source, and the speed the wave travels (speed of light for electromagnetic waves).

Stationary Radar Doppler

Police traffic radar emits an unmodulated continuous wave (CW) and measures' reflections (echoes). Reflections are frequency shifted (Doppler shift) if the target is moving; the faster the target is traveling, the more the frequency shifts. A target traveling toward the radar shifts the frequency higher while a target traveling away from the radar shifts the frequency lower (compared to transmit frequency). The radar, by design, simultaneously transmits a continuous signal while receiving continuous signal echoes.

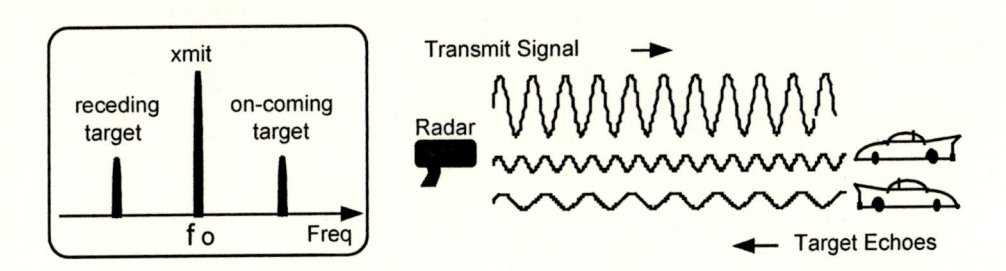

Figure 2.1-1—**Radar Signal**

Target echo frequency (f_t) is a function of radar transmit frequency (f_o) and radar Doppler shift (f_d). A Doppler shift occurs only if the target is moving.
f_d is positive (+) for approaching targets and negative (-) for receding targets.

Approaching Targets
$$f_t = f_o + f_d$$

Receding Targets
$$f_t = f_o - f_d$$

Radar Doppler shift frequency is a function of radar transmit frequency (f_o), speed of wave (c = speed of light), and target velocity (v_t). Note, v_t is positive (+) for approaching targets and negative (-) for receding targets.

$$f_d = \pm 2v_t f_o / c$$
$$v_t = \pm c f_d / 2f_o$$

Donald S. Sawicki

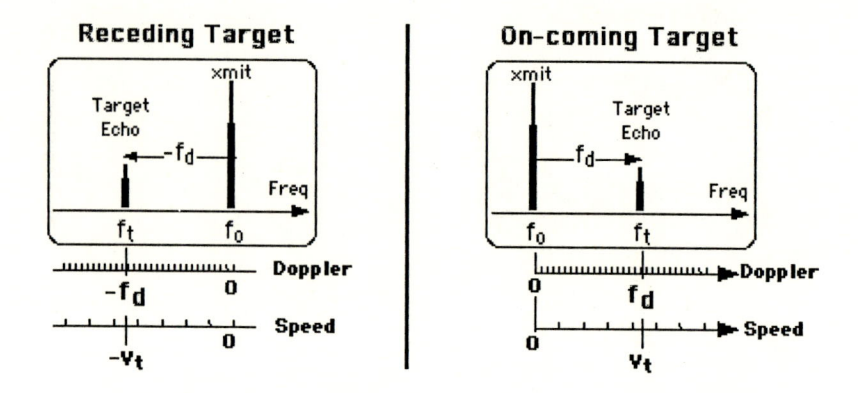

Figure 2.1-2—Stationary Radar Spectrum

$$f_t = f_o \pm 2v_t f_o/c$$

Approaching (on-coming) targets have a positive Doppler shift (target echo higher frequency than transmit); receding (going away) targets have a negative Doppler shift (target echo lower frequency than transmit). If the velocity term v_t is positive (+) target is approaching the radar; if the velocity term v_t is negative (-) target is receding (going away from) the radar.

Ground echoes are usually the strongest signal return, but since the ground is not moving these echoes are not Doppler shifted (ground returns are at frequency f_o) and may be drowned out by transmitter leakage. With moving-mode radar the ground echo is frequency shifted by the speed of the patrol car.

Moving-mode Radar Doppler

Moving-mode radar is slightly more complicated. The target echo frequency is shifted by the relative speed between the target and radar. Target relative speed (to radar) is the sum of target and patrol car speed for opposite direction targets. For same-lane (direction) targets relative target speed is the difference between target and patrol car speed.

**Moving-mode radar depends on two
measurements to derive target speed:**

(1) GROUND ECHO—measures patrol car speed,
(2) TARGET ECHO—measures relative (to radar) target speed.

Traffic Radar Handbook
A Comprehensive Guide to Speed Measuring Systems

Ground echoes are Doppler shifted by the patrol car velocity. The radar tracks the ground echo to determine patrol car (radar) velocity. The radar uses patrol car velocity and relative (to radar) target speed (target echo) to calculate actual target speed.

Receding Target
Rear Antenna

On-coming Target
Front Antenna

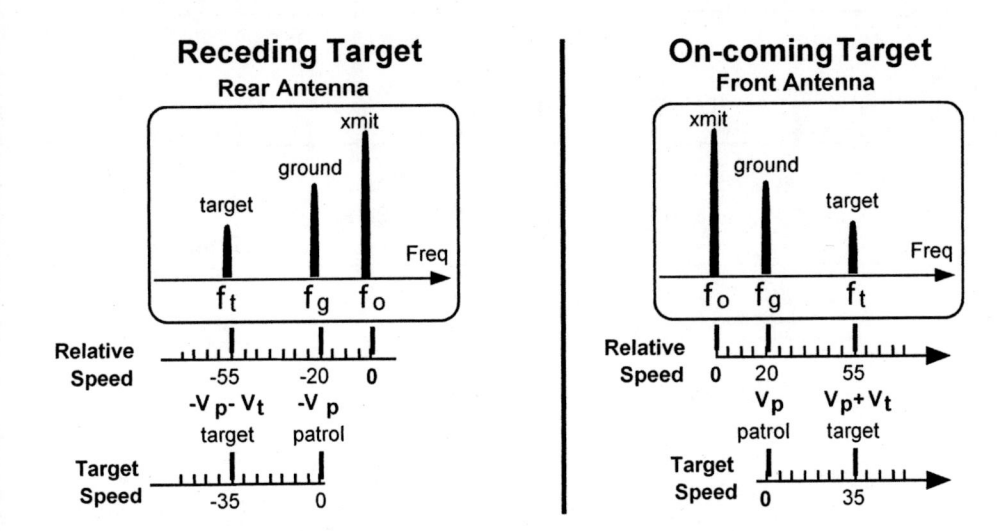

Figure 2.1-3—**Moving Mode Spectrum Opposite Direction Target**

Target Relative Speed to Radar is $V_{relative} = V_p + V_t$
$$V_t = V_{relative} - V_p$$

Donald S. Sawicki

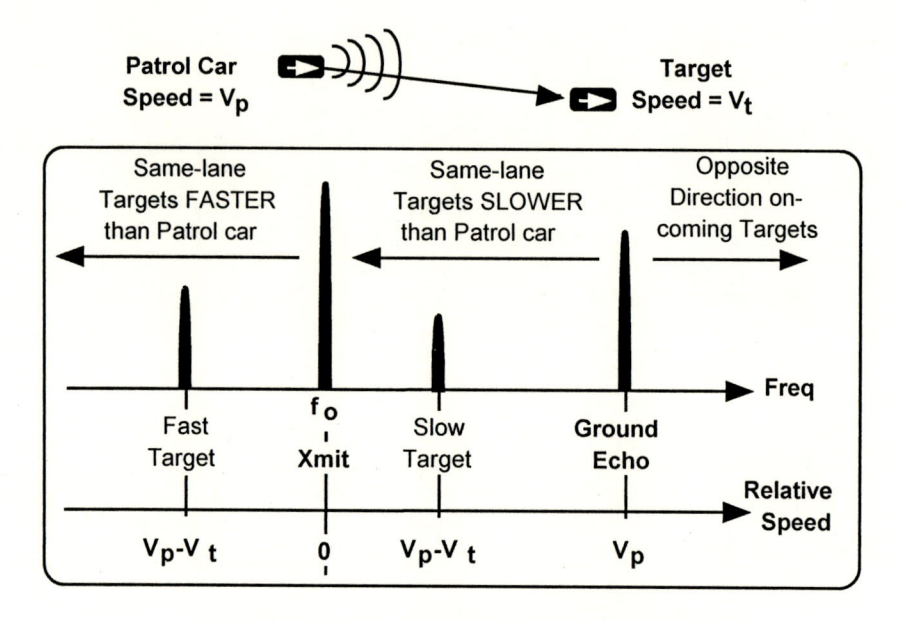

Figure 2.1-4—**Moving Mode Spectrum**
Same Direction (lane) Target Front Antenna

Target Relative Speed to Radar is $V_{relative} = V_p - V_t$

$$V_t = V_p - V_{relative}$$

Note that on-coming (opposite direction) targets have a negative speed (compared to same-lane targets). This type of radar can (if built-in) distinguish between same-lane targets and opposite direction targets.

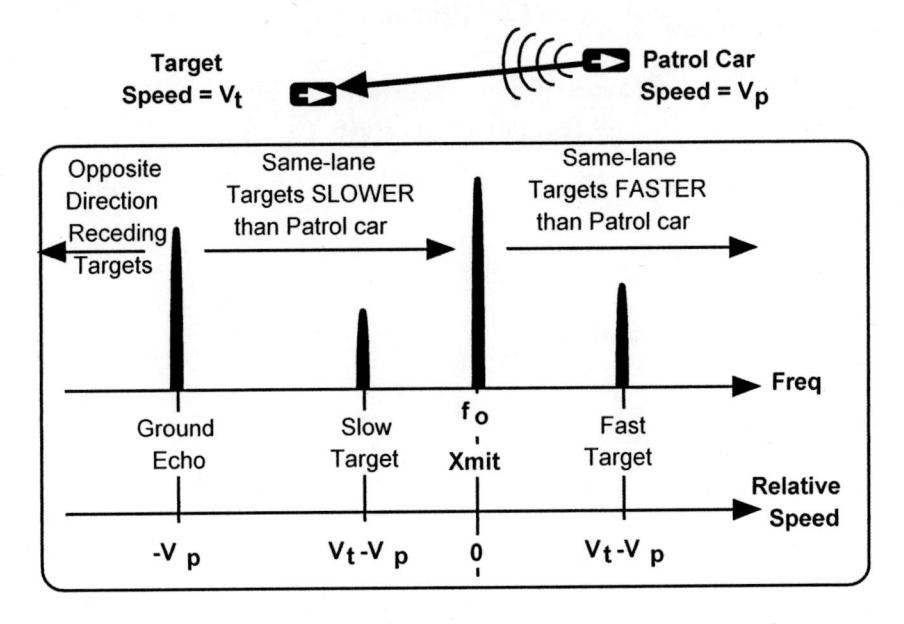

```
                    Target                          Patrol Car
                 Speed = Vt                        Speed = Vp

        ┌─────────────────────────────────────────────────────┐
        │ Opposite     Same-lane          Same-lane           │
        │ Direction    Targets SLOWER     Targets FASTER      │
        │ Receding     than Patrol car    than Patrol car     │
        │ Targets                                             │
        │                                                     │
        │                                                     │
        │                                                Freq │
        │                         f o                         │
        │  Ground        Slow                Fast             │
        │  Echo          Target   Xmit       Target           │
        │                                              Relative│
        │                                               Speed │
        │   -V p         Vt -V p   0         Vt -V p          │
        └─────────────────────────────────────────────────────┘
```

Figure 2.1-5—**Moving Mode Spectrum
Same Direction (lane) Target Rear Antenna**

Target Relative Speed to Radar is $V_{relative} = V_t - V_p$

$$V_t = V_{relative} + V_p$$

Note that receding (opposite direction) targets have a negative speed (compared to same-lane targets). This type of radar can (if built-in) distinguish between same-lane targets and opposite direction targets.

Also see appendix C—Doppler Equations.
Also see (chapter 5.2—Test / Calibration Before Use) Tuning Forks.

Donald S. Sawicki

Chapter 2.2

Radar Cross Section (RCS)
(target reflectivity)

Outline
- Target Shape
- Reflectivity
- Frequency / Polarization
- Angle
- Examples
- Multiple Targets

Microwave traffic radar determines target speed from the Doppler shift of the reflection echo. Larger targets generally have larger echoes; the larger the echo the greater radar detection range. Radar cross-section (RCS) describes a targets' echo and is a measure of reflectivity or radar size (energy reflected / incident). Most targets scatter most of the incident energy, reflecting only a small portion back to the radar. Energy reflected back to the radar (RCS) has dimensions of area, usually **square meters** (m^2). Do not confuse RCS area with physical surface area; both have the same dimensions (area) but gauge different parameters. Sometimes RCS is stated in dbsm—decibels (dB) with respect to (above or below) one square meter (m^2).

The major factors that determine an object's radar cross-section are physical **size, shape**, and material **reflectivity**; as well as radar transmit signal (**frequency** and **polarization**) and **angle** the signal strikes the target.

Target Shape

Targets that disperse energy (reflect energy in directions other than the radar's receive antenna) tend to have a low RCS. Flat surfaces tend to reflect energy better than contoured shapes. Edges and cavities (with diameters on the order of wavelength) tend to be good reflectors. Target shape can be as important as, or even more important than, physical size; stealth aircraft such as the F-117 light bomber and B-2 bomber are extreme examples of how shape can reduce radar cross section.

Reflectivity

Reflectivity is another factor that influences RCS. Metals tend to be good reflectors while carbon composites, fiberglass, and plastics are much more transparent than reflective. Some later model cars have plastic bumpers and body parts in place's metal was formerly used; the less reflective plastic tends to make the vehicle a smaller radar target.

Reflective materials (conductors)—all energy reflected.

All metals (and alloys) are good reflectors. Metal screens with gap spacing less than one quarter (1/4) wavelength reflect as much energy (all) as if the screen were solid metal (no gaps).

BAND	APPROX. 1/4 WAVELENGTH
X	1/4 inch = 7 mm
K	1/8 inch = 3 mm
Ka	1/11 inch = 2 mm

Signals will penetrate (and reflect) a screen with gaps greater than 1/4 wavelength, the greater the gap spacing the more signal penetration and the smaller the reflection.

Non-Reflective materials (insulators)—most energy propagates, some reflected. All materials reflect some energy, most energy propagates with slight beam bending and loss after passing through material.

Frequency / Polarization

Target RCS is also dependent on radar transmit frequency and polarization. Different frequencies and/or polarizations will produce different RCS patterns for the same target. Even ground returns vary with frequency and/or polarization. (See appendix B—Electromagnetic Waves for more details about frequency and polarization). Usually, but not always, the higher the frequency the larger a target's RCS.

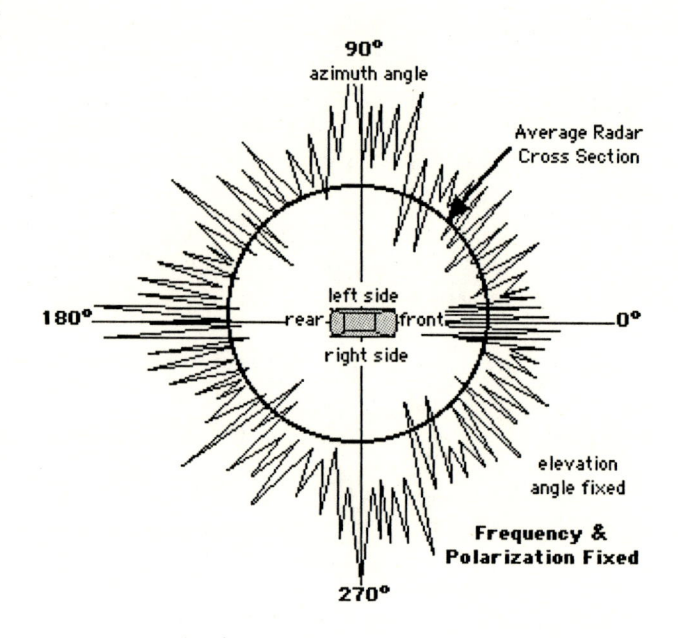

Figure 2.2-1—**Radar Cross Section Plot**

Angle

Most targets (spheres are an exception) reflect different amounts of energy for different azimuth (horizontal) and elevation (vertical) angles. Many times when a target's RCS is stated it will be an average of all important angles. The above figure illustrates how a generic automobile's RCS might vary with azimuth for a fixed elevation angle and a specific frequency and polarization. Generally the front of a car has a smaller RCS than the rear.

A radar detector may increase radar target detection range because antennas (like on a detector) tend to be good reflectors (especially when tuned for radar frequencies). If radar and detector beams are not aligned, or target already has a large RCS, a detector probability will not increase maximum radar target detection. For a radar detector to increase target RCS the detector and radar beams (antennas) must be aligned (narrow radar beam aligned to wide detector beam), and the detector antenna RCS must be large compared to other reflectors (headlights, mirrors, bumpers, grills and metal surfaces). Some reports indicate some jammer antennas increase radar target detection range 10 to 30 percent in some instances.

Examples

Traffic Radar Handbook
A Comprehensive Guide to Speed Measuring Systems

At microwave frequencies average RCS can vary from 0.00001 m² for an insect to over 200 m² for a for a pickup truck. Depending on frequency a man measures about 1 - 4 m² (I wonder whom they measured it on...). Weather RCS (very frequency dependent) varies from about 100 - 10,000 m²; the ground (and clutter) varies from about 1,000 - 100,000 m².

Usually for a class of target (vehicle, fixed wing aircraft, boat, etc.), the larger the target the larger the RCS. Cars and trucks (**vehicles**) tend to have relatively large RCS's (for physical size) compared to other target classes (aircraft, boats, etc.)

	Square Meters (m²)
the Ground and Ground Clutter	1,000 - 100,000
weather (frequency dependent)	100 - 10,000
pickup truck	200
car	100
Jumbo jet	100
large jet airliner or bomber	40
medium jet airliner or bomber	20
large fighter	6
4 passenger jet/small fighter	2
small single engine aircraft	1
conventional winged missile	0.5
cabin cruiser	10
small pleasure boat	2
small open boat	0.02
large bird	0.01
insect	0.00001

Table 2.2-1—**Typical Average RCS at Microwave Frequencies**

Source: Introduction to Radar Systems, Second Edition, Merril I. Skolnik, page 44, 1980.

Usually, but not always, a vehicle's front and rear RCS are different (different shape / reflectors). Also note as radar frequency increases, target RCS (usually) increases.

10 GHz   35 GHz

Donald S. Sawicki

			X band		Ka band	
Year	Make	Model	front	rear	front	rear
1970	Dodge	Pickup Truck	200	300	84	670
1972	Dodge	Dart	100	100	290	110
1970	AMC	Gremlin	200	32	84	73
1972	Oldsmobile	Cutless	100	10	110	73
1965	Ford convertible	Mustang	10	100	55	42
1967	Ford	Cortina	10	10	84	42
1966	Chevrolet	Corvette	10	32	28	28
		bicycle (coasting)	20		2	
		(pedaling)		2		7
		man walking	1	1	3	4
Largest/smallest RCS ratio						
		Truck and Cars	20:1	30:1	10:1	24:1
		Cars only	20:1	10:1	10:1	4:1

Table 2.2-2—**Vehicle Radar Cross Sections (RCS) in m^2**

Source: System Considerations for the Design of Radar Braking Sensors, IEEE Transactions on Vehicular Technology, vol VT-26, no. 2, page 151-160, May 1977.

Note the Corvette has a very low radar cross section (for physical size) because a fiberglass body is much less reflective (to microwaves) then a metal body—not having a metal body is a great way to reduce RCS. The Mustang (convertible) also has a low RCS—not having a metal roof reduces RCS.

The major factors that influence a vehicle's RCS are the area (illuminated) and reflectivity (in direction of antenna). Generally the smaller the illuminated area (physical size), the lower the RCS. A target's RCS is influenced by ALL the reflectors on the target—

- vehicle body,
- radiator / grill,
- bumper,
- head/tail lights / reflectors,
- turn signals / reflectors,
- license plate

- antennae,
- mirrors,
- driver / passengers,
- etc.

The fewer and the smaller the reflectors—the lower the RCS. Note that some **electrically heated windshields**, such as the Ford Instaclear (looks like copper tinted glass) or General Motors (GM) Electriclear, have metal film coatings reflective to microwave signals. These windshields will increase a target's RCS, not to mention prevent drivers from using a radar, or radar detector, from behind the windshield.

TARGET	RCS m^2
Recreational Vehicle (RV)[1]	400
full size Pickup Truck	200
large Car	120
mid size Car	60
small Car	30
Motorcycle[2]	15
Bicycle	8
Man	2

Table 2.2-3—**Estimated Vehicle RCS**
based on above data unless noted.

(1) interpolated from NHTSA 1980 issue paper (see chapter 5.1) range data tracking an Augmented Winnabego Mobile Home.
(2) interpolated by curve fitting surface area to other target's surface areas / estimated RCS's.

For motorist the good news is vehicles with a small RCS are harder to track; the bad news is vehicles with a small RCS may be mistaken for a distant (larger) target. **Motorcycles** are especially vulnerable (as a mistaken target) because the RCS is so small compared to cars and trucks (on the order of 2 to 8 times smaller than a car, 13 times smaller than a pickup truck, and 27 times smaller than an RV), and many more cars and trucks (bigger targets) are on the roads than motorcycles.

Two or more targets in different lanes (and/or with a direct line-of-sight to radar) and traveling at or near the same speed may appear to the radar as a single extended target, with an **RCS greater than the sum** of all target RCS's. Microwave traffic radar uses time and frequency to resolve targets, **multiple targets** in the beam at the same time and at or near the same speed produce an echo return signal that is the vector sum of all targets.

Multiple Targets

A target with a large RCS could have the same signal return power (echo) at a greater range than a close small target. The range a distant large target has the **same (equal) signal return power** as a close small target is a function of both targets' RCS and the range of the close small target

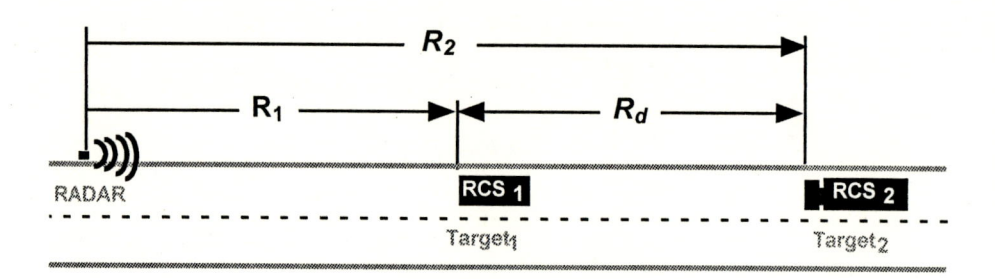

$$R_2 = R_1 (RCS_2/RCS_1)^{1/4}$$
$$R_d = R_1 [(RCS_2/RCS_1)^{1/4} - 1]$$

Figure 2.2-2—**Close Small Target versus Distant Large Target**

The illustration below graphs the range difference (R_d) between 2 targets when the received signals from both targets are equal. The difference is a function of close small target range (R_1) from the radar and the ratio of Target 2 RCS / Target 1 RCS.

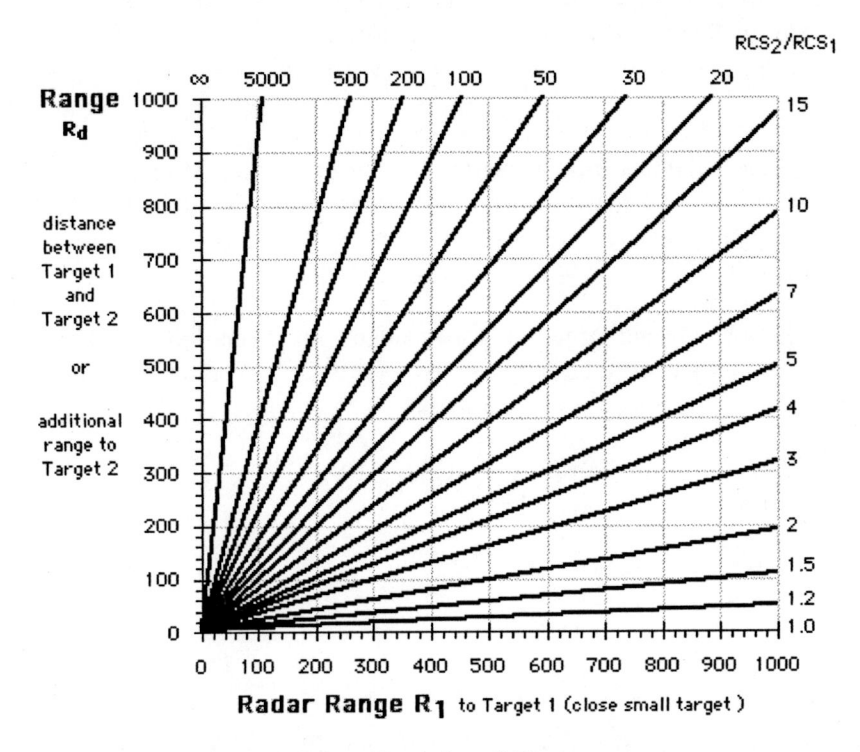

Traffic Radar Handbook
A Comprehensive Guide to Speed Measuring Systems

Figure 2.2-3—**Close Small Target versus Distant Large Target Range Difference**

Both range scales (R_1 and R_d) in the same dimensions (feet, yards, meters, etc.).

EXAMPLEs

- a radar 500 feet from a small car (30 m^2) measures (stronger return signal) a full size car (120 m^2) 210 feet or less behind the small car (710 feet or less from radar).
  RCS 2 / RCS 1 = 120/30 = 4.

- a radar 1000 feet from a small car (30 m^2) measures (stronger return signal) a full size car (120 m^2) 410 feet or less behind the small car (910 feet or less from radar).
  RCS 2 / RCS 1 = 120/30 = 4.

- a radar 500 feet from a small car (30 m^2) measures (stronger

return signal) a pickup truck (200 m²) 300 feet or less behind the small car (800 feet or less from radar).
RCS 2 / RCS 1 = 200/30 = 6.7 or approximately 7.

- a radar 1000 feet from a small car (30 m²) measures (stronger return signal) a pickup truck (200 m²) 600 feet or less behind the small car (1600 feet or less from radar).
RCS 2 / RCS 1 = 200/30 = 6.7 or approximately 7.

In order for a distant large target to maintain equal echo power with a closer small target, the distant large target speed (V_2) must be greater than close small target speed (V_1) by the ratio of RCS's to the 0.25 power.

$$V_2 = V_1 (RCS_2 / RCS_1)^{1/4}$$

EXAMPLEs
$RCS_2 / RCS_1 = 2$, $V_2 = 1.2 V_1$
$RCS_2 / RCS_1 = 3$, $V_2 = 1.3 V_1$
$RCS_2 / RCS_1 = 4$, $V_2 = 1.4 V_1$
$RCS_2 / RCS_1 = 7$, $V_2 = 1.6 V_1$
$RCS_2 / RCS_1 = 13$, $V_2 = 1.9 V_1$
$RCS_2 / RCS_1 = 16$, $V_2 = 2 V_1$ (V_2 twice the speed of V_1)
$RCS_2 / RCS_1 = 50$, $V_2 = 2.7 V_1$

The above equation applies to moving mode radar by using closing (or opening for opposite direction targets) speeds in place of target speeds. Solving above equation for moving radar, if V_p is patrol car speed;

$$V_2 = (V_1 + V_p) (RCS_2 / RCS_1)^{1/4} - V_p$$

Stationary radar $V_p = 0$

Time / Range Distant Target Echo Greater
When the radar starts tracking **APPROACHING** traffic, if $R_2 < R_1 (RCS_2/RCS_1)^{1/4}$ and $V_2 > (V_1+V_p)(RCS_2/RCS_1)^{1/4}-V_p$, then Target 2 will have a greater signal echo than Target 1 entire track time (both targets in the beam). When the radar starts tracking **RECEDING** traffic, if $R_2 < R_1 (RCS_2/RCS_1)^{1/4}$ and $V_2 < (V_1+V_p)(RCS_2/RCS_1)^{1/4}-V_p$, then Target 2 will have a greater signal echo than Target 1 entire track time (both targets in radar detection range).

Traffic Radar Handbook
A Comprehensive Guide to Speed Measuring Systems

Larger distant **approaching** targets traveling slower than (V_1+V_p) $(RCS_2 RCS_1)^{1/4}$ - V_p, but faster than closer target, require shorter range (closer) to maintain an equal or greater echo than the closer target. To determine distant target (Target 2) range (R_2 requires knowing distant target speed and RCS as well as close target (Target 1) range, speed, and RCS, and **radar tracking time**. The faster and/or the larger (RCS) the distant target, the greater the range for the distant target echo to be greater than or equal to the close target echo. Also the shorter the radar tracking time (t), the greater the range of the distant target. In practice time (t) is a multiple of radar sample time, minimum time (t) is 1 radar sample period.

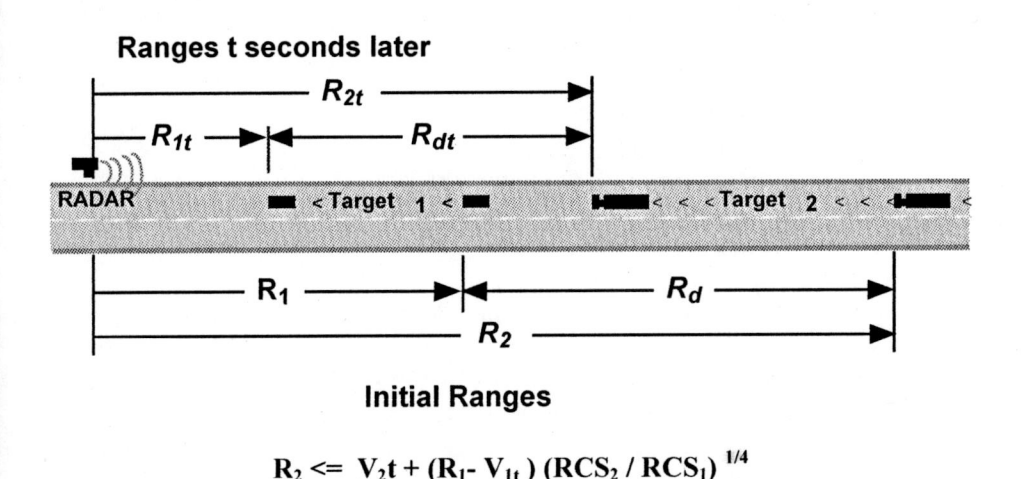

Ranges t seconds later

R_{1t}, R_{2t}, R_{dt}

RADAR < Target 1 < , < Target 2 <

R_1, R_2, R_d

Initial Ranges

$$R_2 <= V_2 t + (R_1 - V_{1t})(RCS_2 / RCS_1)^{1/4}$$

$$RCS_1 < RCS_2$$
$$V_1 < V_2 < V_1 (RCS_2 / RCS_1)^{1/4}$$
t < Target 1 time to even with Radar (R_1 / V_1)
t < time Target 2 even with Target 1 [$R_1 - R_1 (RCS_2 / RCS_1)^{1/4}) / (V_1 - V_2)$]

Figure 2.2-4—**Range Distant Target Echo
Greater than or Equal to Close Target Echo
Approaching (on-coming) Traffic**

On initial track, Target 1 at range R_1 and Target 2 at range R_2 Target 2's echo is greater than Target 1's echo. Time t seconds later, Target 1 at range R_{1t} and Target 2 at range R_{2t}, both target echoes equal.

The above equation applies to **moving mode** radar by using closing speeds in place of target speeds. Solving above equation for moving radar, if V_p is patrol car speed;

$$R_2 < = t(V_2+V_p) - t(V_1+V_p)(RCS_2/RCS_1)^{1/4} + R_1(RCS_2/RCS_1)^{1/4}$$

$$V_1 < V_2 < (V_1+V_p)(RCS_2/RCS_1)^{1/4} - V_p$$
$$t < \text{Target 1 time to even with Radar } [R_1/(V_1+V_p)]$$

GIVEN
- V_2 — Target 2 velocity
- V_1 — Target 1 velocity
- R_1 — Target 1 initial range
- RCS_2 — Target 2 RCS
- RCS_1 — Target 1 RCS
- t — Radar track time
- V_p — patrol car speed (moving mode radar)

Stationary radar $V_p = 0$

Larger distant receding targets traveling slower than $(V_1+V_p)(RCS_2/RCS_1)^{1/4} - V_p$ have a larger signal echo entire time both targets in radar detection range. Receding targets traveling faster than $(V_1+V_p)(RCS_2/RCS_1)^{1/4} - V_p$ require shorter range (closer) to maintain an equal or greater echo than the closer target.

Chapter 3.1

The Basic Radar Gun

Outline
- Detection Range
- Operation
- Controls
- Display / Indicators

Also see chapter 6.3—Laser Radar Operation

The basic microwave traffic radar is an unmodulated CW (continuous wave) transmitter that illuminates a target to measure the Doppler shift of target reflections. Most microwave traffic radars measure speed of one target at a time; range or angle information is not available. When multiple targets are present, the operator has no way to know for sure which target the radar is tracking. Most radars display the target with the strongest reflection (usually, but not always, the closest target); some models have the option to display the fastest (or a faster) target. Some models track and display two targets—the fastest or faster target, and the strongest target.

Detection Range
Traffic radar detection **range** varies with model and unit, and target. Maximum range can be as high as a mile (1.6 km) or more for large targets under good conditions, or as low as a hundred feet (30 m) or less for small targets under bad conditions. Factors that effect detection range include;

- Radar parameters
— transmit power
— antenna gain (size)
— receiver sensitivity
— integration period (sample time); longer period (time), greater range
— frequency

- target Radar Cross Section (RCS)
— varies with size (surface area), frequency, and angle

- noise sources
— near-by transmitters, high voltage lines/transformers, etc.

- atmospheric conditions
— temperature, humidity, weather, etc.

Moisture in the atmosphere (rain, fog and/or humidity) as well as smoke and dust particles attenuates K and Ka band signals more than X band. Detection range is generally greater in the morning than in the afternoon; usually temperature is higher in the afternoon and the sun has had a chance to heat objects increasing background noise. Note that **atmospheric conditions** (temperature, humidity, smoke, and dust particles) do NOT affect radar accuracy, only target detection range.

A phenomena know as **surface ducting** (also referred to as a **ground-based duct**) can increase radar detection range significantly. Surface ducts have a height of about 30-70 feet (10-20 meters) and occur when upper air is exceptionally warm and dry COMPARED to surface air. Most of the time the greater the altitude the lower the temperature, when a layer of relatively constant cooler and more humid air forms near the surface a duct occurs. Conditions favorable to duct formation include heat radiating from the earth such as on cool summer night; cool air from the base of a thunderstorm can also form a duct. Long straight flat stretches of concrete (such as runways and highways) are known for forming microwave surface ducts. Large bodies of water tend to form ducts more often than land masses (water cools surface air).

OPERATION

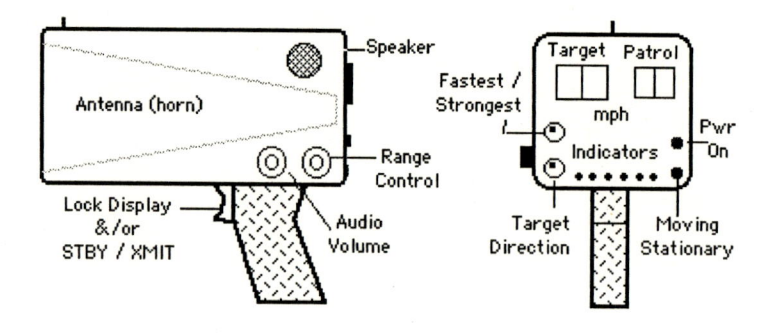

Figure 3.1-1—**Typical Radar Gun**

(not all operations available on all models)

Physical Configurations
Voltage supplied by car battery and/or internal battery

Single unit
—hand-held radar gun (see figure 3.1-1 above)
—patrol car and/or motorcycle mounts (some models)
—Remote Control (some models)

Multi-unit (typical configurations and options)
—PROCESSOR (display and controls) module
—mounted ANTENNA module (with transmitter/receiver)
—2nd rear facing antenna (some models)
—operator selects front, rear, or both antennas for operation
—wired or wireless Remote Control (some models)
—still and/or video camera (photo radar)
—large display to show drivers their speed (safety radar)

Frequency (X, K, or Ka band)
X band: 10.525 GHz ± 25 MHz
K band: 24.150 GHz ± 100 MHz, or 24.125 GHz ± 100 MHz
Ka band: 33.4 - 36 GHz—operate on single frequency (± 100 MHz)
 in a13 channel band, and/or as wideband.

Wideband Radar: (some Ka Band radars)
 Intended to defeat radar detectors. Wideband radars (**Ka** band) operate on a single fixed frequency (operator selects one of several available), and/or in a frequency hop mode. A frequency hop mode radar dwells on a fixed frequency for a fraction of a second (on the order of 1/10 of a second or so) and hops to another frequency. The radar cycles between a number of different frequencies.

Transmission Time
 Continuous: Many radars continually transmit and process every target when the radar is turned on. Some radars only display targets above a preset speed (audio alert setting switch).

Instant On (some models)
Intended to defeat radar detectors. Instant On radars allow the operator to control radar transmission; the operator transmits only after selecting a target, and only long enough to get a speed reading.

Pulsed (some models)
Intended to defeat radar detectors. Some pulsed radars only transmit periodically on the order of every several seconds or so, and then only transmit long enough (100 ms or so) to get a speed measurement. Some pulsed radars transmit only one short burst on operator command. Pulsed radars, by design, do not establish a tracking history.

Radar Mode(s)
   Stationary—Many traffic radars operate from a fixed position only—stationary radar. Many stationary radars measure on-coming (approaching) traffic only; some measure on-coming and/or going (receding) traffic.
   Moving Mode (Moving Mode Radar)—Most moving mode radars can operate in either stationary or moving mode; a few older radar models are moving mode only. Most also display patrol car speed, a few older models do not. Many moving radars measure on-coming targets only, some measure on-coming and/or receding traffic. Some moving radars can also measure same-lane traffic in front of (forward looking antenna) and/or behind (rear looking antenna) the patrol car.
   Target Direction (some models)—All stationary radars measure on-coming (approaching) traffic, some measure on-coming OR going (receding) traffic, and some radars measure on-coming AND going traffic (and some of these indicate target direction—on-coming or going).
   Most moving mode radars measure opposite direction on-coming targets (front antenna). Some radars measure opposite direction and/or same-lane targets. Moving radars with a rear antenna measure same-lane traffic and/or opposite direction (receding) traffic. Many radars that measure targets in more than one direction usually have a display that indicates target direction. Moving radar requires a minimum patrol car speed (typically a few to 20 mph); same-lane mode also requires a minimum speed difference between target and patrol car (typically a few to 5 mph).
   The table below summarizes radar type and common configurations (varies with model) of radar mode (stationary / moving) with respect to target direction and front or rear antenna.

| | Stationary mode || Moving-mode ||
Antenna	Front	Rear	Front	Rear
Stationary Radar	on-coming and/or going	on-coming and/or going	-	-
Moving Radar	"	"	on-coming and/or same-lane	same-lane and/or going

Table 3.1-1—**Target Direction and Radar**

SIDE NOTE: A stationary radar (or moving mode radar set to function in a stationary position) operated from a moving patrol car (and pointed in the direction of travel—front) will measure the patrol car speed. The ground is a huge target that swamps out discrete target echoes such as from other vehicles (moving or not). The radar processes the ground echo, Doppler shifted by patrol car speed, as a target. If the radar is angled (off patrol car direction), the displayed speed reads low (see chapter 4—cosine effect error).

Antenna Front / Rear (some models)—Multi-piece single antenna radars usually mount the antenna pointed in the forward direction of the patrol car; some mount the antenna pointed in the rear. Some systems have two antennas, forward and rear, and the operator selects one or both antennas for use. Antennas can be mounted inside the patrol car (behind a windshield), or outside on the side of the vehicle between the roof and hood.

Lock Target Speed
Most radars can lock (freeze) displayed target speed. Some radars have a separate display for locked speed so the radar can continue to measure other targets.
  Manual Lock: operator manually locks target speed. (most models)
  Auto-Lock: freezes target display (or LOCKED display) and/or photographs or video tapes automatically when a preset target speed has been met or exceeded. Only unattended photo/video radars or older models have this feature. Some courts will not accept the radar as evidence if this function is used.

Fastest Target option (some models)
Most radars display the speed of the target with the strongest echo (usually the closest target); some radars have the option to display the fastest (or a faster) target, not necessarily the strongest echo. In fastest (or faster) target mode the

Donald S. Sawicki

radar may require longer tracking time (on the order of seconds instead of fractions of a second). Some models track and display the fastest and strongest targets in the radar beam.

Audio Tones (most models)
—audio tone equal or proportional to target Doppler frequency. (Audio Doppler)
—alert tone when a target has met or exceeded a preset speed.

Self-test (most models)
Self-test transmit frequency and/or some portion of radar electronics. Also see chapter 5.2—Test / Calibration Before Use.

Accuracy (typical)

Stationary Radar
± 1 mph

Moving Radar
target speed: ± 2 mph
patrol car speed: ± 1 mph

Traffic Radar Handbook
A Comprehensive Guide to Speed Measuring Systems

CONTROLS
(not all functions on all models)

Transmit / Standby (Instant On Radar)
    Transmit—radar transmitting
    Pulse—brief transmission (0.1 to 1 second or more)
    Standby—radar on but not transmitting

Manual-Lock (most models)
    lock or release target speed displayed

Auto Lock
    —Photo / Video Radar
    —some older models

Stationary / Moving Mode (Moving Mode Radar)
    Stationary—radar operates from a stationary position
    Moving mode—radar operates from a moving patrol car

On-coming / Going / Both (stationary mode)
    On-coming—radar measures approaching targets only
    Going—radar measures receding (going) targets only
    Both—radar measures approaching and receding targets

Opposite and/or **Same-lane Traffic** (Some Moving Mode Radars)
    FRONT ANTENNA
    Opposite direction (on-coming traffic) and/or
    Same-lane (direction) traffic
    REAR ANTENNA
    Same-lane (direction) traffic and/or
    Opposite direction (going traffic)

Front / Rear / Both Antenna(s) (some models)

Fastest / Strongest Target (some models)
    Largest—display strongest target echo (usually the closest)
    Fastest—display fastest target echo
    (some models track and display both largest and fastest targets)

Range Control—radar receiver sensitivity threshold setting

Audio Squelch—squelch target Doppler audio tone

Audio Alert
Set target speed to sound alert (when a target meets or exceeds the preset speed)

Audio Volume—target and alert tones

Test (most models)
DC Power
—may transmit when DC power applied (not Instant On radars)
—many run self-test on power-up

DISPLAYS / INDICATORS
LEDs (Light Emitting Diodes) and/or LCD (Liquid Crystal Display) (not all indicators/functions on all models)

```
        Fastest                          Fastest
         or                                or
Target  Locked                  Target   Locked   Patrol
±000    000                     ±000     000      00

TRAFFIC DIRECTION               TRAFFIC DIRECTION
+ on-coming, - going            + opposite direction, - same-lane

Stationary Radar                    Moving Radar
```

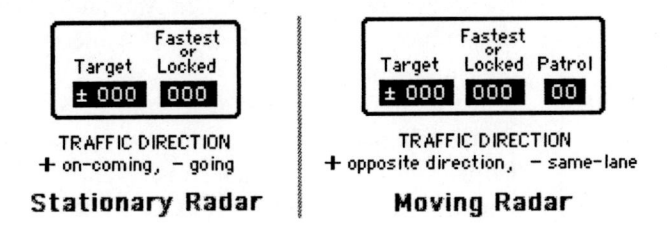

Figure 3.1-2—**Speed Display Configurations**

Display—Target Speed
  —has a minimum and maximum speed measuring range
  —mph, kmh, or knots (some models operator selects units)
  —Target Direction indicates approaching or receding target (not all models)
  —LOCK switch freezes this display if no separate display for locked speeds
  —Target Direction (not all models) for moving radar (single antenna) indicates opposite or same-lane (direction) traffic; for stationary radar with forward and rear antennae indicates on-coming (approaching) or going (receding) traffic.

Display—Locked (target) Speed (not all models)
  Allows operator to lock a target speed and continue to track targets.
  Some models use this display for Fastest / Faster Target.

  NOTE: radars that track two targets (strongest and fastest/faster) generally use the middle display for the fastest target and the target display for the strongest target.

Display—Patrol Car Speed (Moving mode radar)
  usually smaller speed measuring range than target display

Speaker (most models)
  —audio tone equal or proportional to target Doppler frequency
  —some models sound alert tone when target exceeds preset speed

Donald S. Sawicki

Indicators (typical)
TRANSMIT	—	radar transmitting
STANDBY —		radar on but not transmitting
LOCK —		displayed speed frozen
RFI —		Radio Frequency Interference Another radar, transmitter, or jammer signal detected. Most models inhibit speed measurements
XMIT ERROR	—	transmit frequency out of tolerance, Some models test frequency automatically and/or when operator initiated
BIT ERROR	—	Built-in-test error radar test (automatically and/or operator initiated) portion of electronics.
LOW VOLT	—	battery voltage below specification. operation inhibited, some models display warning.
PWR ON/OFF	—	Radar ON Many units run self-test (several seconds) on power-up. May indicate radar transmitting.

Traffic Radar Handbook
A Comprehensive Guide to Speed Measuring Systems

Chapter 4.1

Cosine Error
Geometry
Microwave and Laser Radars

Outline
- Setup
- Overpass
- Hills / Curves
- Cosine Error

Setup

Figure 4.1-1—**Cosine Effect Setup**

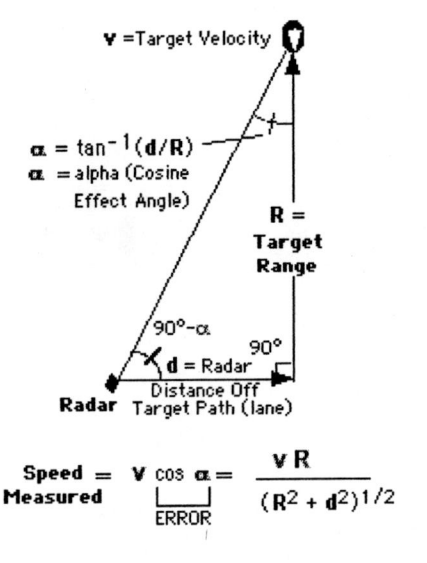

Traffic radars measure the relative speed a target is approaching (or receding) the radar. If a target is traveling directly (collision course) at the radar, the relative speed is actual target speed. If the target is not traveling directly toward (or away) the radar but slightly off to avoid a collision, the relative speed with respect to the radar is slightly lower than target speed. The phenomenon is called the Cosine Effect because the measured speed is directly related to the cosine of the angle (**alpha**) between the radar and **target direction** of travel (see figure).

V = Target Velocity
$\alpha = \tan^{-1}(d/R)$
α = alpha (Cosine Effect Angle)
R = Target Range
d = Radar Distance Off Target Path (lane)

$$\text{Speed Measured} = V \cos \alpha = \frac{VR}{(R^2 + d^2)^{1/2}}$$

ERROR

The cosine effect angle (alpha) is the angle between the radar and the target direction of travel. Target range from radar and radar distance off the road (really the distance between radar and the point the target would be closest to radar if target continues in same direction) determine the cosine effect angle. Note that the **road direction and antenna direction (direction antenna pointed) are completely irrelevant**, only the angle (alpha) matters (radar stationary).

Donald S. Sawicki

Antenna direction (alignment to patrol car direction) is important in **moving mode** radar. A mis-aligned antenna measures target speed high if the misalignment is great enough, the target is approaching the moving radar, and the target is traveling slower than the radar (chapter 4.3—Cosine Effect on Moving Radar).

As long as the angle (alpha) remains relatively small, the error (cosine of alpha) is tolerable. The larger the angle, the larger the error and the lower the displayed (relative) speed. On a straight section of road, **radar distance from the road** and the **range** of the target determine the **angle**. The greater the distance the radar is off the road and/or the closer the target, the larger the angle (and error). When the target is even with the radar (alpha equals 90 degrees) the target speed, with respect (relative) to the radar, is zero.

The Cosine Effect applies to both **microwave** radars and **laser** radars (ladars) as well as to targets traveling in any direction (on-coming or going traffic at any angle). Most traffic radars do not account for the Cosine Effect; across the road microwave radars (such as photo radars) are an exception. These systems point the beam at a known fixed angle across the road and compensate the measured target speed for the Cosine Effect.

Overpass

Figure 4.1-2—Cosine Effect from an Overpass

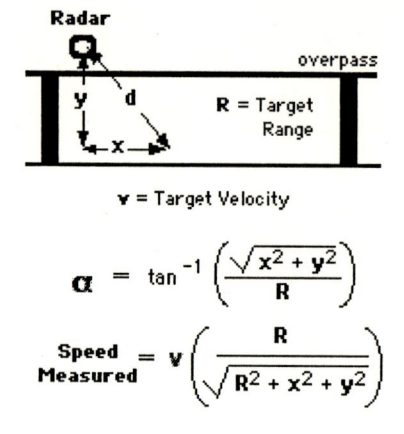

R = Target Range

v = Target Velocity

$$\alpha = \tan^{-1}\left(\frac{\sqrt{x^2 + y^2}}{R}\right)$$

$$\text{Speed Measured} = v\left(\frac{R}{\sqrt{R^2 + x^2 + y^2}}\right)$$

The radar distance from (off) the road is the line-of-sight distance from the radar to the road (target path). If the radar is on an overpass (shooting cars running under the overpass) or hill for example, the radar distance from the road is the distance from the radar position to the road (target path) as in the side figure. In the figure the traffic is traveling directly away or into the page.

Traffic Radar Handbook
A Comprehensive Guide to Speed Measuring Systems

In the side figure, **x** represents the horizontal distance and **y** represents the vertical distance from the road to the radar. The line-of-sight distance is **d**. If either the horizontal or vertical component is zero, the equations reduce to that shown in figure 4.1-1—where **d = y** (if **x** = 0), or **d = x** (if **y** = 0). When applying the equations, all distances must use the same unit dimensions (feet, meters, etc.). When calculating the angle (alpha) using the inverse tangent function (arctan), the unit dimension of the calculated angle is radians (rad), not degrees.

Hills / Curves

On hills or curves target direction (with respect to radar) is changing, this causes the Cosine Effect angle (alpha) to change. A changing Cosine Effect angle results in measured target speed changing, the faster the angle changes the faster measured target speed changes (acceleration or deceleration component). If measured speed changes too fast the radar misses (does not display) target speed.

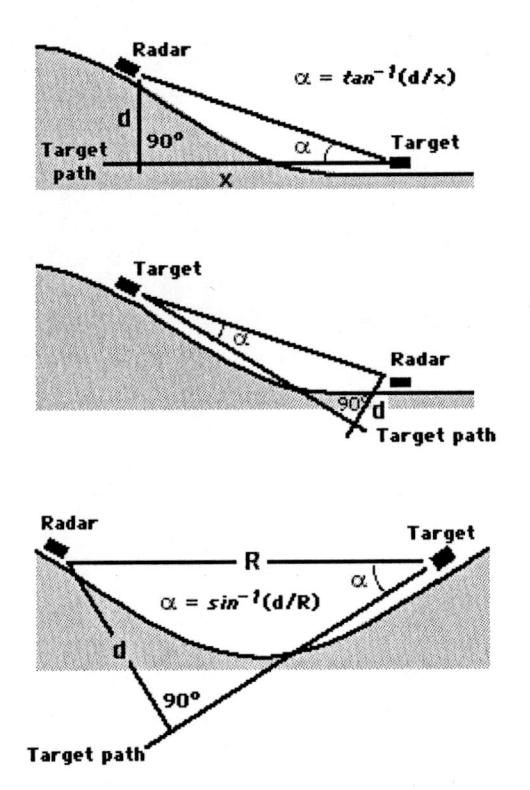

Figure 4.1-3—**Cosine Effect Due to Hills / Curves**

Donald S. Sawicki

Note the above figures illustrates targets on a hill (side view), or a curved road (top view). Alpha is the Cosine Effect angle, d is radar (closest) distance from target path. The steeper the hill or the tighter the curve the greater the angle alpha, and the greater the measured speed error and the greater the acceleration component. Moving radar introduces another component that generally increases the target acceleration component for approaching or receding targets, and decreases for same-lane targets.

Also see;
chapter 4.5—Targets on a Curve
chapter 5.4—Operational problems / Target Acceleration

Cosine Error

The below figure is a graphical representation of the Cosine Effect for measured speed, as a **percentage of true speed** versus **angle** (alpha) between radar and target—the larger the angle the larger the error and the lower the measured target speed. For example at angles of only a few degrees the measured speed is 99 to 100 percent of actual; at an angle of 60 degrees the measured speed is half (50 percent) the actual target speed.

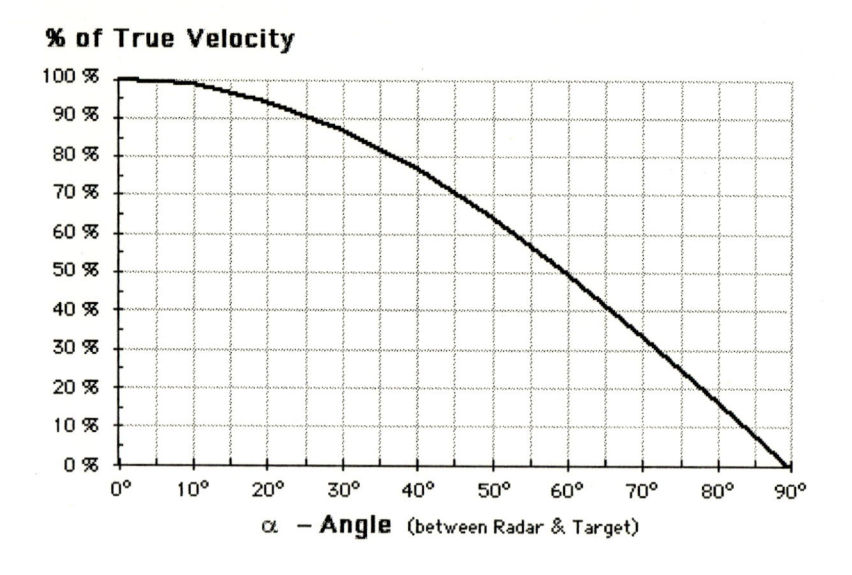

Figure 4.1-4—**Angle vs Measured Speed (percent of actual speed)**

Traffic Radar Handbook
A Comprehensive Guide to Speed Measuring Systems

Chapter 4.2

Cosine Effect
on Measured Speed
Microwave and Laser Radars

The Cosine Effect angle (alpha) changes with time for a moving target; the rate of change dependents on target range and speed as well as radar distance off the road. The figure below graphs measured target speed versus time (seconds) on a straight road for targets traveling at 30, 50, and 75 mph and radar (stationary) distance from (off) the road of 50 feet. Radar display update time shown in the figure is one second.

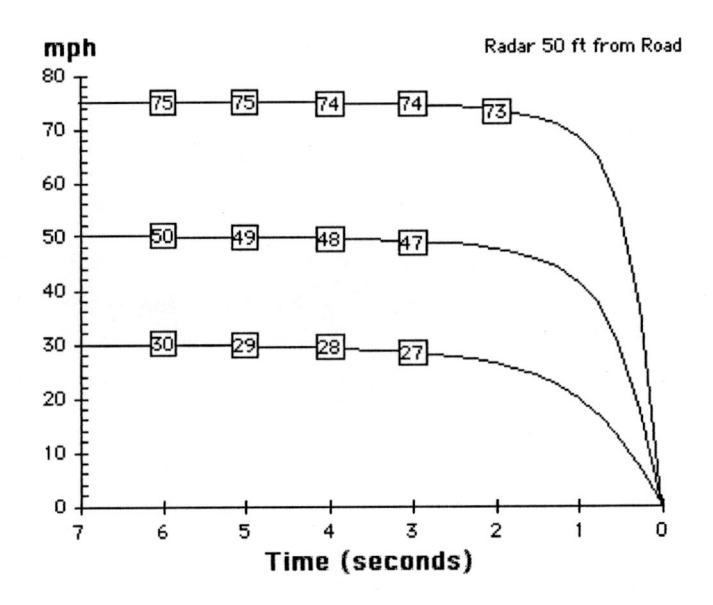

Figure 4.2-1—**Cosine Effect and Radar Speed Reading**

The closer a target is to the radar the larger the measured speed error (the lower the measured speed). Depending on target speed and radar position off the road, the last several seconds before a target passes the radar the measured target speed has a relative acceleration (due to the Cosine Effect) greater than 1 mph

43

Donald S. Sawicki

per second. Many radars will not display target speed if relative speed (measured) changes faster than about 1 mph (or kmh) per radar sample time (integration or cycle time), typically about 1/4 to 2 seconds.

The figure below graphs target range when the measured speed error is **-0.5 mph** (-0.8 kmh) and relative target acceleration is **± 1 mph/second** (1.6 kmh/second) as a function of radar distance off the road and target speed. The figure applies to both approaching and receding targets. The acceleration component gets larger (+) for approaching targets and smaller (-) for receding targets. Time scales (right side) equate true target speed with target time (seconds) and range (feet) to radar.

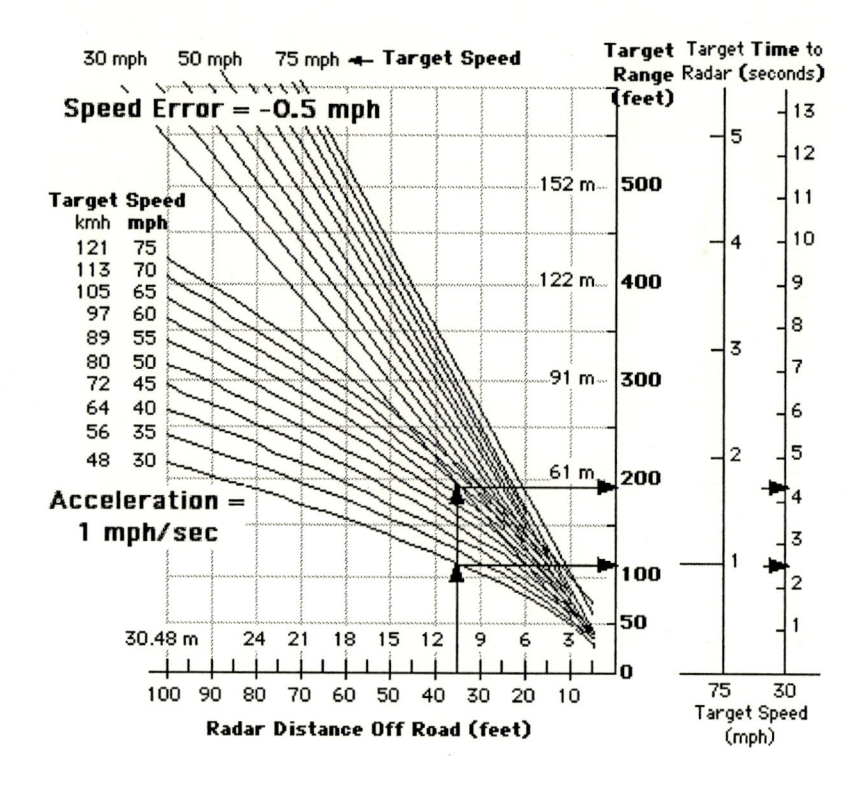

Figure 4.2-2—**Target Range/Time vs Cosine Errors**

For example, when a target traveling at 30 mph is tracked by a radar 35 feet from target direction of travel (radar distance off the road), the radar measures target speed low by 0.5 mph (30 - 0.5 = 29.5 mph) at a distance of 189 feet, about 4.3 seconds before target even with radar. Target acceleration of 1 mph per second occurs at 112 feet away from the radar, about 2.5 seconds before target even with radar. The closer the target to the radar, the lower the measured target velocity and the greater the acceleration component—when the target is even with the radar the measured target speed is zero.

Radar off road: 35 ft, Target Speed: 30 mph

	Range	Time to even
Speed error -0.5 mph	189 ft	4.3 sec
Acceleration 1 mph/sec	112 ft	2.5 sec

Minimum Range

Radar minimum target range (based on acceleration due to Cosine Error) depends on the angle between the radar and target (**the Cosine Effect**), the time (radar **sample time**) the radar requires to establish a track history (typically tens of seconds to one second or more), and radar tolerance to **accelerating or decelerating targets**. The greater the radar distance from the road and/or the longer the sample time, the greater the minimum operating range.

Radars typically specify an accuracy of ± 1 mph. For a radar to maintain this accuracy the radar should only track targets with a constant speed within about plus or minus 1 mph (or kmh) during a sample period (radar integration or cycle time). Sample periods vary with model and range from 0.25 seconds (250 ms) or less to 2 seconds or more. Laser radars typically have a sample period of 0.3 seconds (300 ms) or more.

Minimum radar operating range can be estimated from radar accuracy (typically ± 1 mph) and sample time by determining when the Cosine angle effects target acceleration by more than 1 mph or kmh (the accuracy) during a sample time. A radar with a 250 millisecond (0.25 seconds) sample time will have difficulty tracking a target changing speed by more than 1 mph per 250 ms (4 mph/sec) or 1 kmh per 250 ms (4 kmh/sec).

Donald S. Sawicki

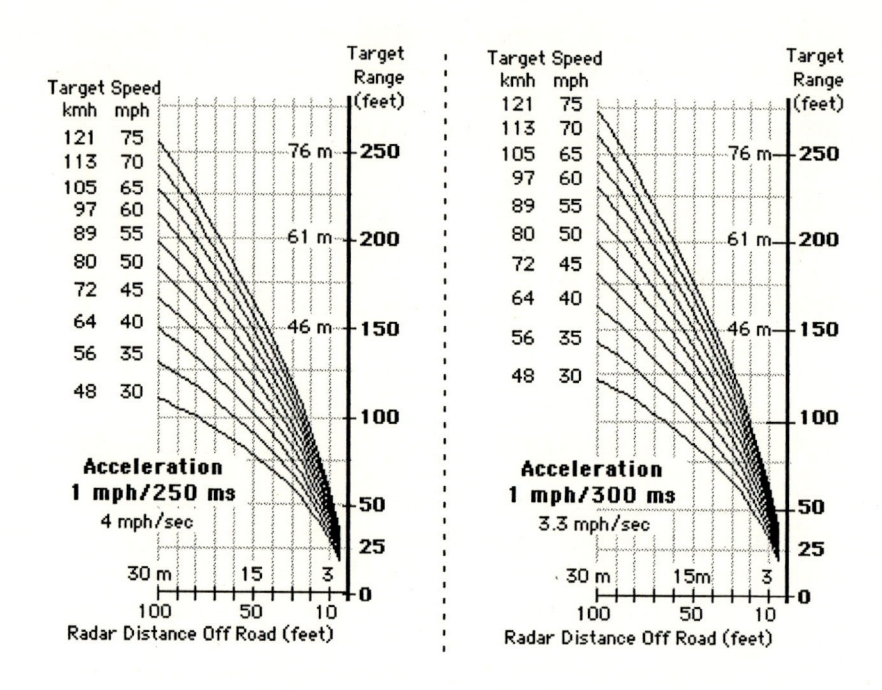

Figure 4.2-3—
Cosine Effect and Relative Target Acceleration

For example when a radar is 50 feet from the road a 30 mph target has an acceleration (really deceleration which is a negative acceleration) component of 1 mph/0.25 seconds due to the Cosine Effect at about 79 feet from the radar; a 70 mph target has an acceleration component of 1 mph/0.25 sec at about 157 feet. The figure above illustrates acceleration for a 1 mph per 250 and 300 milliseconds sample time (1.6 kmh per 250 and 300 ms).

Minimum Range due to Radar Sample Time

In situations the radar is very close to the road (target path) the limiting factor for minimum range may be the radar sample time only (not any Cosine Effects). Most radars require a good target track for at least one period (sample time) to compute target speed. Radar sample time and target speed are the primary factors influencing minimum range.

To compute Minimum Range in **feet** from radar sample time (**milliseconds**), and target speed (**mph**);

Min Range = (**22 / 15000**) x (Sample Time) x (Target Speed)

Traffic Radar Handbook
A Comprehensive Guide to Speed Measuring Systems

For example; a 300 ms (sample time) radar tracking a 55 mph target has a minimum range (to track target at least one full cycle) of 24 feet ((22/15000) x 300 x 55).

ms	30	35	40	45	50	55	60	65	70	75	mph target
100	4	5	6	7	7	8	9	10	10	11	feet
200	9	10	12	13	15	16	18	19	21	22	
300	13	15	18	20	22	24	26	29	31	33	
400	18	21	23	26	29	32	35	38	41	44	
500	22	26	29	33	37	40	44	48	51	55	
600	26	31	35	40	44	48	53	57	62	66	
700	31	36	41	46	51	56	62	67	72	77	
800	35	41	47	53	59	65	70	76	82	88	
900	40	46	53	59	66	73	79	86	92	99	
1000	44	51	59	66	73	81	88	95	103	110	

MINIMUM RANGE (feet) GIVEN TARGET SPEED (mph) AND RADAR SAMPLE TIME (milliseconds—ms).

To compute Minimum Range in **meters** from radar sample time **(milliseconds)**, and target speed **(kmh)**;

Min Range = **(1 / 3600)** x (Sample Time) x (Target Speed)

For example; a 300 ms (sample time) radar tracking a 110 kmh target has a minimum range (to track target at least one full cycle) of 9 meters ((1/3600) x 300 x 110).

ms	50	60	70	80	90	100	110	120	130	140	kmh target
100	1	2	2	2	3	3	3	3	4	4	meters
200	3	3	4	4	5	6	6	7	7	8	
300	4	5	6	7	8	8	9	10	11	12	
400	6	7	8	9	10	11	12	13	14	16	
500	7	8	10	11	13	14	15	17	18	19	
600	8	10	12	13	15	17	18	20	22	23	
700	10	12	14	16	18	19	21	23	25	27	

Donald S. Sawicki

800	11	13	16	18	20	22	24	27	29	31
900	13	15	18	20	23	25	28	30	33	35
1000	14	17	19	22	25	28	31	33	36	39

MINIMUM RANGE (meters) GIVEN TARGET SPEED (kmh) AND RADAR SAMPLE TIME (milliseconds—ms).

Traffic Radar Handbook
A Comprehensive Guide to Speed Measuring Systems

Chapter 4.3

Cosine Effect on Moving Radar
Microwave Traffic Radars

Operation

With moving-mode radar the target echo frequency is shifted by the relative speed between the target and radar. Target relative speed (to radar) is the sum of target and patrol car speed for opposite direction targets. For same-lane (direction) targets relative target speed is the difference between target and patrol car speed. The radar must also track the ground echo (speed of ground) to determine patrol car (radar) speed. The radar uses patrol car speed and relative (to radar) target speed (target echo) to calculate actual target speed. Any error in measured patrol car speed (ground echo) or relative target speed (target echo) translates directly to computed target speed (error).

Ground echoes off angle (not directly in front of direction of travel) cause the patrol speed to measure low. Moving targets (same direction and slower than the radar) mistaken by the radar as the ground echo also compute patrol speed low. Targets traveling faster than the radar should not be mistaken for a ground echo because the Doppler shift is negative. Patrol speed errors due to off angle ground echoes, or a slow moving vehicle mistaken for ground echo, cause a LOW measured patrol speed (**shadowing**). When patrol speed measures low the error favors the target (low target speed reading) for same-lane (direction) targets; and works against the target (high target speed reading) for opposite direction targets.

For opposite direction targets if the radar-target Cosine Effect error is negligible a low patrol speed reading translates directly into a high (computed) target speed. As the radar target Cosine angle increases relative target speed decreases and tends to cancel out any (shadowing) low patrol speed readings (high target speed readings).

Donald S. Sawicki

Figure 4.3-1—
Moving Mode Radar Variables

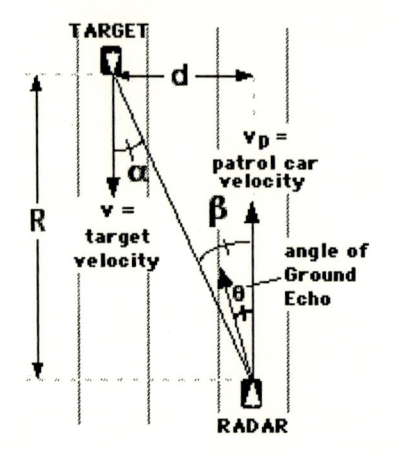

Parameters that affect moving radar cosine errors;
—Patrol Car Speed (Vp)
—Radar Angle to Patrol Car direction (theta)
—Target Speed (V)
—Range (R)
—Distance between Paths (lanes) (d)

The angles (alpha and beta) between the radar and target with respect to directions of travel (patrol car and target) are generally the predominant error sources (and depend on **R** and **d**). The larger the angles, the greater the error (the lower the measured target speed). If the patrol car and target directions of travel are parallel alpha equals beta. In most situations target velocity (V) and patrol car velocity (Vp) have minor affects on the cosine error unless the antenna is mis-aligned.

If the radar is mis-aligned (**theta** not equal to 0 degrees) the angle theta between the radar (patrol car) direction of travel and ground echoes can be a significant factor under certain conditions. Ground echoes from moving radar are affected by cosine error, the greater the error the lower the measured patrol car speed. For approaching targets a lower patrol car speed translates into a higher target speed (target speed = closing speed - patrol car speed).

Target speed will only measure higher than true speed when the target is approaching the patrol car AND the cosine angles (alpha and beta) between radar and target are small (typically less than 5 degrees) AND the angle (theta) between patrol car and ground echo is large (typically greater than 5 degrees). Patrol car and target speeds (Vp and V) are significant, **patrol car speed greater than target speed** increases the error (the greater the difference the larger the error and the higher the measured target speed).

The figure below graphs measured target speed error (percentage above true target speed) for patrol car and target approaching each other. The angles (alpha and beta) between radar and target are both zero (worse case). The target speed error is a function of angle (theta) between radar and ground echo and the ratio of patrol car speed (Vp) to target speed(V). The ratio is 1 (one) when patrol car speed equals target speed (both vehicles approaching each other at the same speed).

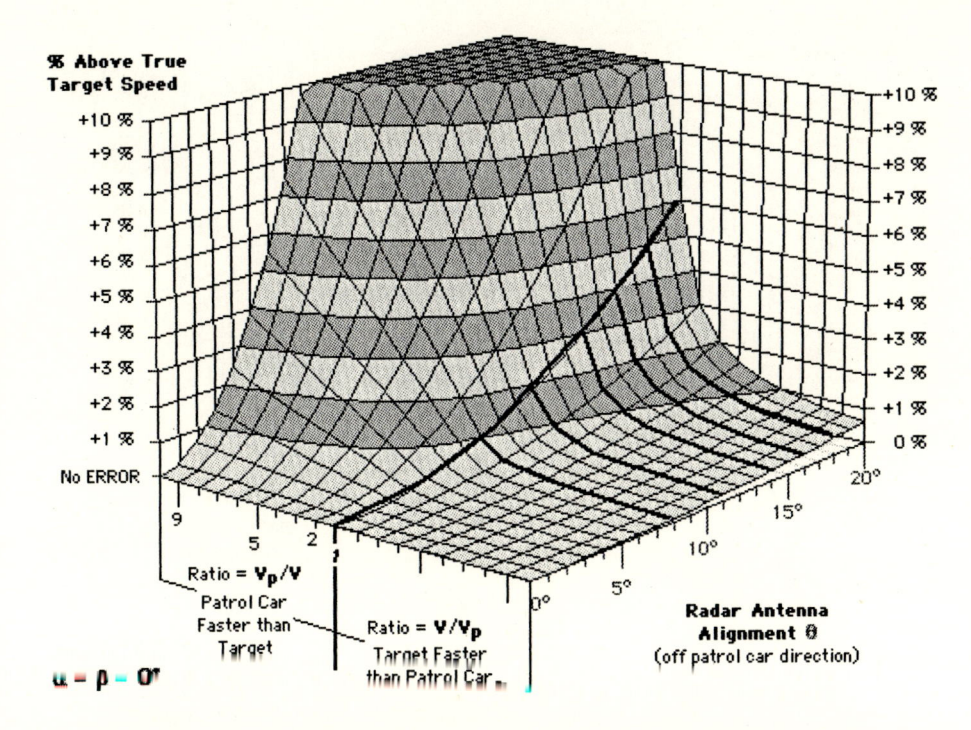

Traffic Radar Handbook
A Comprehensive Guide to Speed Measuring Systems

Figure 4.3-2—**Target Speed Error (percent above true speed)**

The target and patrol car speed ratio axis is divided into two sections; patrol car faster than target (Vp/V) and target faster than patrol car (V/Vp). The scale (number) is the multiplication factor, for example a target traveling twice as fast as the patrol car (V/Vp) has a factor of 2 (right side of scale). A patrol car traveling twice as fast as target has a factor (Vp/V) of 2 (left side of the axis).

As the angle between radar and target increases, the cosine error due to ground echoes becomes less significant. The cosine error due to angle between target and radar tends to nullify the cosine error due to ground echo cosine errors. The figure below graphs measured target speed error for an angle between radar and target of 5° (as opposed to 0 degrees in previous figure).

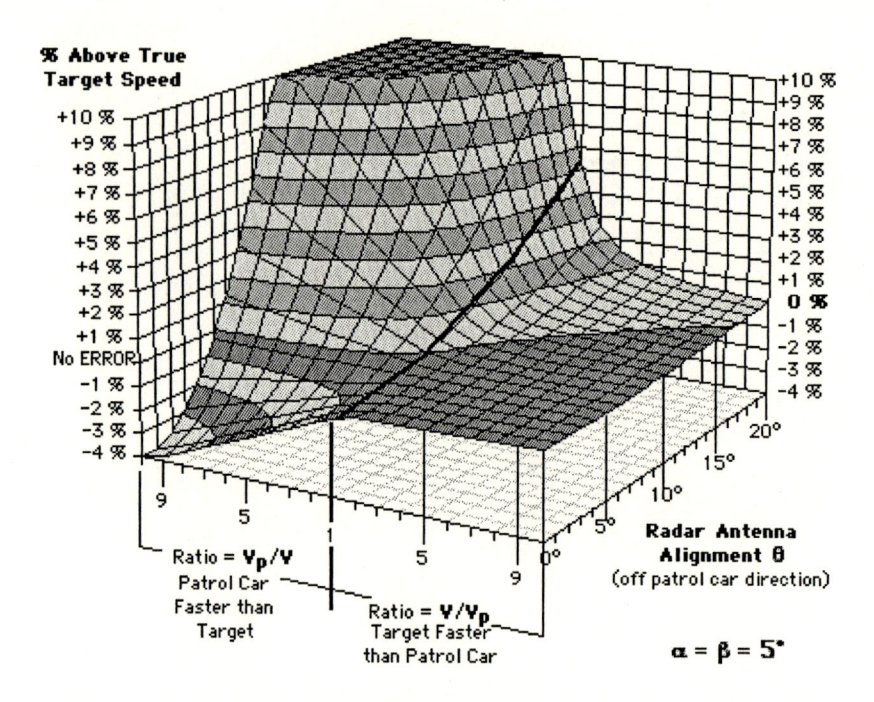

Figure 4.3-3—**Measured Target Speed Error (alpha = beta = 5°)**

To minimize patrol car speed (cosine) error the radar antenna should align with patrol car direction (front and/or rear facing antennas). If the radar antenna is properly aligned, the graph in figure 4.2-2—Target Distance/Time vs Cosine Error applies to moving mode radar with the following additions and modifications;

—target and patrol car directions are parallel,
—all target speeds become closing/opening speeds,
—distance off road becomes distance between paths,
—target acceleration error becomes closing/opening acceleration error,
—target speed error becomes closing/opening speed error.

Summary

Three major factors influence moving mode radar Cosine Errors;

- patrol car speed compared to target speed,
- radar alignment to patrol car direction of travel,
- distance between radar and target(s) paths.

To minimize moving radar Cosine Error against opposite direction targets the patrol car speed should be less than target speed (see above figures), and radar antenna (front and/or rear) alignment should be within a few degrees of direction of travel (of patrol car). Also the less distance between patrol car and target paths, the smaller the Cosine Error. Note that this last factor (also) influences stationary radar, and the others do not.

For Same-Lane Radar (target same direction as radar), target speed compared to patrol speed has little effect if radar antenna aligned to patrol car direction AND target in exact same lane, not a lane or two over, as patrol car (alpha = beta = theta = 0°). Targets one or more lanes over should be traveling faster than patrol car to prevent an excessively low speed readings.

Donald S. Sawicki

Chapter 4.4

Cosine Effect on Across the Road Radar
Microwave Traffic Radars

Across the road photo and safety radars generally operate at relatively short range and have narrow horizontal beams. The horizontal beamwidth (beta) and alignment angle (alpha) determine the extent of the Cosine Effect error—the larger the beamwidth the more the Cosine Effect error varies (between beam edges); the greater the alignment angle the larger the Cosine Effect error.

Target time in the beam (time radar has to measure target) depends on target speed, radar beamwidth and alignment (Cosine Effect angle), radar distance from the road (really target path), and target width and length. Target (6 x 15 ft) time in a 5 degree beam angled at 22 degrees can vary from 0.16 seconds (radar 5 ft or 1.5 m from road, and a 150 mph or 240 kmh target) to 2.4 seconds (radar 60 ft or 18 m from road, and a 20 mph or 30 kmh target).

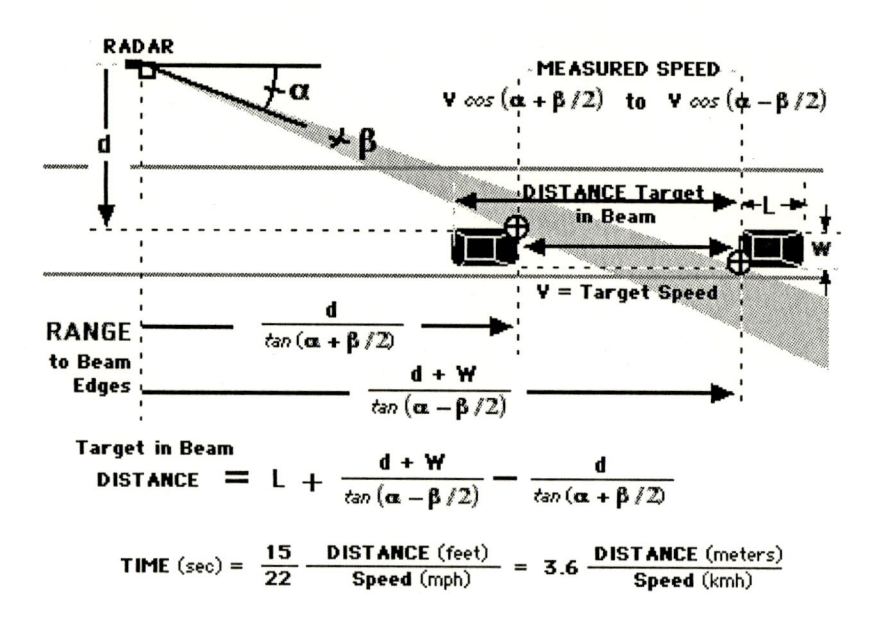

MEASURED SPEED $V \cos(\alpha + \beta/2)$ to $V \cos(\alpha - \beta/2)$

V = Target Speed

RANGE to Beam Edges:
$$\frac{d}{\tan(\alpha + \beta/2)}$$
$$\frac{d + W}{\tan(\alpha - \beta/2)}$$

Target in Beam
$$\text{DISTANCE} = L + \frac{d + W}{\tan(\alpha - \beta/2)} - \frac{d}{\tan(\alpha + \beta/2)}$$

$$\text{TIME (sec)} = \frac{15}{22} \frac{\text{DISTANCE (feet)}}{\text{Speed (mph)}} = 3.6 \frac{\text{DISTANCE (meters)}}{\text{Speed (kmh)}}$$

Figure 4.4-1—Cosine Effect on Across the Road Radar

If alpha = 22° and beta = 5°,
DISTANCE in Beam = 0.630 d + 2.824 W + L

If alpha = 22.5° and beta = 5°,
DISTANCE in Beam = 0.603 d + 2.747 W + L

A radar may not be able to measure a target's speed if the target is traveling fast enough to pass through the beam in less than (faster than) 1 radar integration cycle.

Donald S. Sawicki

Chapter 4.5

Targets on a Curve
Microwave and Laser Radars

A target traveling in a straight line at a constant speed does not have an acceleration component—acceleration is zero. However a target traveling on a curve at a constant speed does have a radical acceleration component due to the action of turning. A curve also causes measured target speed to read low due to the Cosine Effect angle. The Cosine Effect error is a function of where on the curve the target is (with respect to the radar)—the deeper into the curve (the larger the Cosine Effect Angle) the greater the error (the lower the measured speed). Radars (microwave and laser) generally cannot successfully measure targets on a curve (or in a turn) due to relative (to radar) target acceleration and/or Cosine Effect angle.

V = TARGET SPEED
α = Cosine Effect Angle
V_m = MEASURED TARGET SPEED
   = V cos α

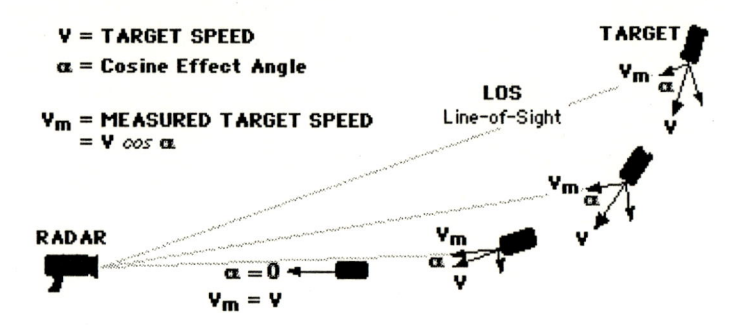

LOS Line-of-Sight

RADAR

α = 0
V_m = V

Figure 4.5-1—**Target on a Curve**

Target velocity has magnitude and direction—the magnitude is target speed (V), direction is tangential to the curve (perpendicular to radical acceleration). The radical acceleration component has magnitude and direction—magnitude (A) is velocity squared divided by curve radius, direction is aligned to the center of the curve radius.

Traffic Radar Handbook
A Comprehensive Guide to Speed Measuring Systems

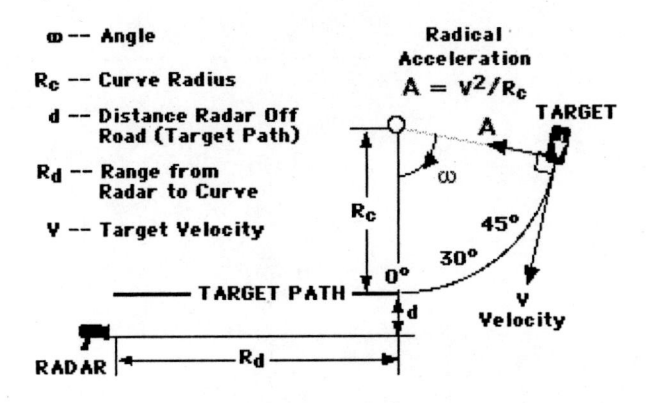

```
ω -- Angle                        Radical
                                  Acceleration
Rc -- Curve Radius                A = V²/Rc
d  -- Distance Radar Off                        TARGET
      Road (Target Path)
Rd -- Range from
      Radar to Curve
V  -- Target Velocity                    45°
                                    30°
          ――― TARGET PATH ―  0°
                                         V
                                         Velocity
    RADAR  ←――― Rd ―――→
```

Figure 4.5-2—**Target Parameters on a Curve**

d shown positive (+), if radar on other side of path d is negative (-).

Radar measured target speed (V_m) and acceleration (A_m) are a function of curve radius (R_c), angle (omega) on the curve, target speed (V), radar range from curve (R_d) when omega equals 0°, and radar distance off target path (d). The faster the target, the tighter the curve, the deeper into the curve, and/or the closer the radar—the greater the relative (to radar) acceleration and the lower the measured target speed. Typically at about omega equal 15° relative (to radar) acceleration and the velocity error start becoming significant factors affecting a traffic radar's ability to track and accurately measure targets.

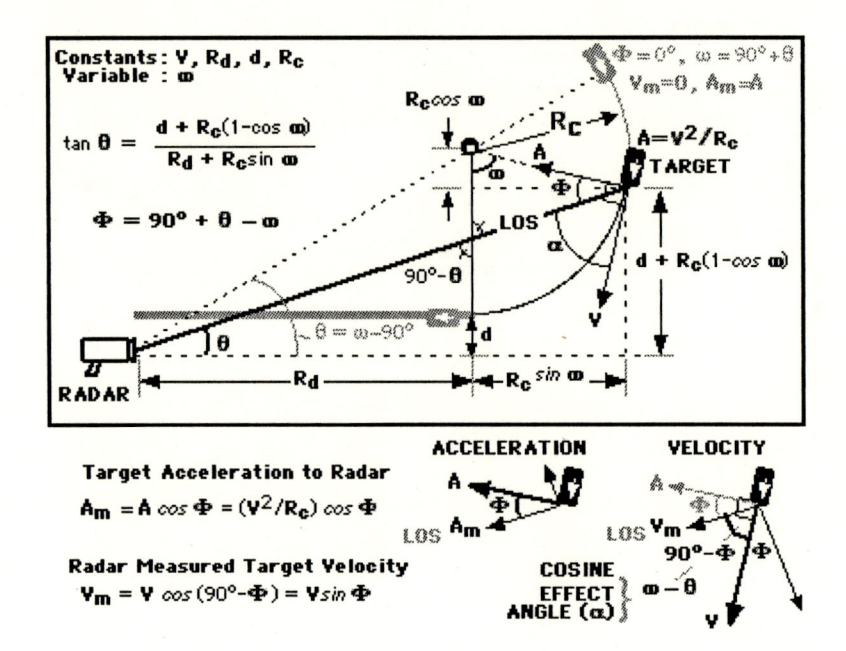

Figure 4.5-3—**Velocity and Acceleration Vectors on a Curve**

―――――――――――――――――――――EXAMPLES―――――――――――――――――――
d=0

CURVE RADIUS	TARGET CURVE ANGLE	RADAR RANGE to CURVE	TARGET SPEED	MEASURED (relative) SPEED	MEASURED (relative) ACCELERATION
50 ft	3 deg	100 ft	25 mph	25 mph	0.95 mph/sec
		1000 ft		25 mph	0.96 mph/sec
100 ft	7 deg	100 ft		24.8 mph	1.06 mph/sec
		1000 ft		24.8 mph	1.11 mph/sec
200 ft	15 deg	100 ft		24.4 mph	0.99 mph/sec
		1000 ft		24.2 mph	1.16 mph/sec
300 ft	27 deg	100 ft		23.6 mph	1.00 mph/sec
		1000 ft		22.6 mph	1 31 mph/sec
400 ft	41 deg	100 ft		22.5 mph	1.00 mph/sec
		1000 ft		20.1.mph	1.36 mph/sec

Donald S. Sawicki

Chapter 5.1

NHTSA 1980 Tests
National Highway Traffic
Safety Administration

Outline
- Test
  —Problem Summary
  —Variations Between Radars
  —Beamwidth, ERP, and Low Voltage Summary
- Recommendations

This section summarizes the **Police Traffic Radar ISSUE PAPER** (DOT HS-805 254)—U.S. Department of Transportation (DOT) - National Highway Traffic Safety Administration (NHTSA) dated February 1980. The entire report is available on the Internet at the Office of Law Enforcement Standards (OLES) web site (http://www.eeel.nist.gov/oles/index.html).

TEST

Seven radar units (six different models) from four different manufacturers were tested for problems. The study listed several recommendations to address problems encountered with some of the radars in the test. All radar units tested operate at X band; two radars were handheld (the CMI radars), all the others (5) were multi-piece radars.

Manufacturer	Model	Test Radar*
CMI	**Speedgun 6**	D (handheld)
CMI	**Speedgun 8**	E (handheld)
Decatur Electronics	**MV-715**	C
Kustom Signals	**MR-7**	G
Kustom Signals	**MR-9**	A
M.P.H. Industries	**K-55**	B and F

* Unofficial sources (not NHTSA)

Traffic Radar Handbook
A Comprehensive Guide to Speed Measuring Systems

The test data does not identify the exact radar model used but instead categorizes the radar units as A, B, C, D, E, F, and G (**7 radars tested, 6 different models** from 4 manufacturers). Note: * indicates from unofficial sources (not NHTSA). The test revealed several operational (chapter 5.4) and multiple interference (chapter 5.5) problems. Below is a summary of the test results.

RADARS AFFECTED	%	AFFECTED BY
A B C D E F	86%	patrol speed shadowing (moving mode)
A B D E F G	86%	internal 100 watt mobile radio
A B C D F G	86%	internal CB transmissions
A B C F G	71%	panning error (all units that were 2 piece)
B C D E F	71%	Mechanical Interference, (scanning error)
A B E F G	71%	internal 2 watt hand-held radio
B C D E	57%	internal air-conditioner and heater fans
B D F	43%	speed bumping (moving mode)
A B F	43%	external police radios transmitting 20-0 feet away
C D	29%	ignition and alternator
F	14%	external CB transmissions
B D F G	57%	TUNING FORK(s) mis-labeled or missing

5 out of 12 (over 40 percent) of the tuning forks were mislabeled (mis-calibrated) by 1 mph; four were labeled low by 1 mph, one was high by 1 mph.

All radars tested transmitted in band except for one—radar F transmitted slightly out of X band (10.500-10.550 GHz) by 900 kHz (10.5509 GHz) and off center (10.525 GHz) frequency by 25.9 MHz.

Other moving mode abnormalities;
40 mph radar (B) picked-up 37 mph Pinto then lost for several seconds before measuring another car way behind Pinto, 41 mph radar (D) tracked 57 mph truck well behind 35 mph Pinto before tracking Pinto, 41 mph radar (G) picked up 57 mph truck 100 yards (90 meters) behind 33 mph Pinto.

Variations (max and min) Between Radars Tested
Beamwidth: 13.3 to 24.6 degrees

Donald S. Sawicki

    Effective Radiated Power (ERP): 26.3 to 134 milliwatts (14 to 21 dBm)
        Max Detection Range: 105 feet to 1.1 miles (32 m to 1770 km)
        Low Voltage Condition: 6 to 11.9 Vdc

MAX DETECTION RANGE TEST: Two stationary radar detection range test (1 radar against Ford Pinto on 2 runs) conducted, and 86 moving mode radar test runs (radar and target approaching each other). Moving mode tested each radar against 3 different targets—2 door Ford Thunderbird, Augmented Winnabego Mobile Home, and 2 door Ford Pinto.

Beamwidth, ERP, and Low Voltage Summary

Radar	A	B	C	D	E	F	G	med	avg
Beamwidth	13.3°	20.4°	17.5°	18.8°	18.6°	24.6°	14.3°	18.6°	18.2°
ERP (mW)	85	134	120	34.25	39.2	26.3	56	56	71
Low Volt (V)	11.9	6.0	7.6	10.2	10.3	6.6	11.9	10.2	9.2

Note, Test Radars B and F were both MPH K-55 radars; Radar B measured beamwidth was 20.4 degrees, Radar F measured beamwidth was 24.6 degrees.
Also see Web site (http://CopRadar.com) chapter 3.1 (TEST DATA sub-links).

RECOMMENDATIONS

Based on the test results the report suggested several recommendations as summarized below;
    —Adopt radar standards and require police agencies to follow standards,
    —Develop policy guidelines for radar maintenance, testing, and calibration,
    —Keep adequate maintenance and calibration records,
    —Establish minimum training standards,
    —Develop State-level certification (renew every 1 to 3 years),
    —Develop radar workshops and seminars for traffic adjudication personnel,
    —Establish State-level policy/procedural guidelines to ensure proper use.

NHTSA recommended standards, based on the 1980 test, may be found under the title: Performance Standards for Speed Measuring Devices, United States Department of Transportation and National Highway Traffic Safety Administration, Federal Register, Volume 40, Number 5, January 8, 1981. Below is a list of some of the recommended specifications.

Performance Characteristics		Minimum Requirement
Frequency	X Band	10.525 GHz +/- 25 MHz
	K Band	24.150 GHz +/- 100 MHz
Beamwidth	X Band	18 degrees
	K Band	15 degrees
Minimum Detection Range		500 feet (150 meters)
Accuracy	Stationary	+/- 1 mph
	Moving Mode	+/- 2 mph
		+/- 1 mph (patrol car)
Tuning Fork Tolerance		+/- 1/2 percent

Donald S. Sawicki

Chapter 5.2

Test / Calibration
Before Use

Outline
- Operator Testing
- Self-test (internal calibration)
    — Transmit Frequency
    — Low Voltage
- Tuning Forks
- Lab / Shop Calibration

Operator Testing

To insure radar hardware is setup and functioning properly the operator should;

- run radar **self-test** (most models have a self-test)
    — before and after use (before / after shift) at a minimum
    (some radars automatically self-test on power-up and/or periodically)

- check radar with calibrated **tuning fork**
    — several forks (several speeds) makes for a better check
    — minimum of 2 forks (2 speeds) required to check moving mode radar

- check radar against target of known speed (several speeds better)
    — not always practical for moving mode radar

- Insure radar **lab / shop calibration** current
    — radar should be labeled with lab/shop calibration data (BY, DATE, DUE)
    — most states require traffic radars to be lab/shop calibrated 1 or 2 times/year

Also see Tests (chapter 6.5—Operational Problems with Laser Radar) for operator tests before laser radar (ladar/lidar) use.

Self-test (internal calibration)
The degree a radar self-test and/or automatically adjust itself varies with model and ranges from none to testing and/or adjusting 50 percent or more of the electronics. Because self-test (if any) only checks/adjusts a portion of the radar electronics, the radar should be checked against an actual target of known speed (preferably several speeds), and with calibrated tuning forks.

Radar self-test (if any) runs;
- on power-up
- when the operator initiates (pushes a button)
- automatically (periodically on the order of minutes)
- automatically before and/or after a target speed is locked
- any or all of the above

Tests / Automatic Adjustments
may include (but not limited to);
- transmit frequency
- supply voltage (car or internal battery)
- a test target signal
- portions of the digital circuits
- portions of the analog circuits
- display indicators
- any or all of the above

TRANSMIT FREQUENCY

To get an accurate speed measurement the transmit frequency must operate in specific limits. A radar transmitting out of specification on the high side may result in a measured speed that is high; transmitting out of specification on the low side may result in a measured speed that is low. The speed error depends on designed center frequency, actual transmit frequency, and target speed. To maintain a ± 1 mph (or kmh) accuracy most radars (not some older models) inhibit operation if transmit frequency drifts out of specification (± 25 MHz for X band, ± 100 MHz for K and Ka band radars). Some later models automatically adjust transmit frequency periodically and/or during self-test; if the frequency cannot be adjusted to specifications operation is inhibited.

Radar measured speed (including error), based on transmit frequency (f_x) and designed transmit center frequency (f_o), is predictable. The radar Doppler filters measure target speed and are a function of designed center transmit frequency. Note that c represents the speed of light.

Donald S. Sawicki

- Target Radar Doppler shift $f_d = 2v\, f_x / c$
- Radar Doppler filters (V_m) based on center frequency f_o
$V_m = c\, f_d / 2\, f_o$

$$v_m = v(f_x / f_o)$$

V_m —speed measured by radar
V—target speed
f_x —transmitted frequency (Hz)
f_o —center band frequency (Hz),

$$\text{Speed Error (\%)} = 100\, (1 - f_x / f_o)\, \%$$

LOW VOLTAGE

Some radars are sensitive to operating supply voltage (car battery or internal battery). Radar electronics can usually handle an excessively high voltage with simple regulator circuits, but the problem of low voltage is not so easy to manage. Low voltage can cause a radar to malfunction—false readings, erred readings, or simply blank out. A low voltage condition (varies with model) may be a problem if the radar does not monitor supply voltage and indicate if voltage too low and/or inhibit operation. Many states will not certify (for police use) a radar that does not indicate a low voltage condition and/or inhibit operation.

Tuning Forks

Tuning forks are used to quick check radar speed accuracy. A radar can measure a vibrating tuning fork and produce a speed reading proportional to the resonant frequency of the fork (fork resonance frequency equals radar Doppler shift).

The displayed speed is a function of radar transmit frequency and tuning fork resonance frequency.

$$f_d = 2\, v_t\, f_o / c$$

$$v_t = c\, f_d / 2\, f_o$$

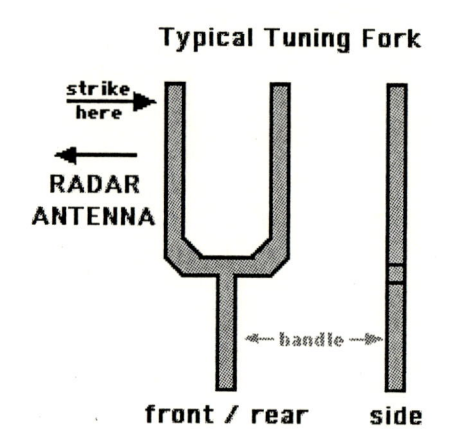

F_d = Radar Doppler Shift (Hz) = Tuning Fork Resonance (Hz)
V_t = Target Velocity
F_o = Transmit Frequency (Hz)
c = speed of light (same units as V_t)

To check a microwave radar with a tuning fork the radar must be transmitting and the fork must be vibrating. To start the fork vibrating gently strike the top **side** (not front or rear, see above figure) against a hard object such as wood or plastic (not metal). Once the fork is vibrating place it within a few inches (within 10 centimeters) in front of the radar antenna, make such the **side** (not front or rear) is facing (pointing at) the antenna.

One tuning fork does not necessarily indicate how accurate the entire speed measuring range (20 to 90 mph for example) is functioning; multiple tuning forks are required to test the entire range. **Moving-mode** radar requires a minimum of **two tuning forks** for test, one to simulate a ground return (patrol car speed) and the other to simulate target speed.

Moving mode radars exposed to two vibrating forks with different frequencies should display the difference in frequency between the forks as target speed, and the lower frequency fork as the displayed patrol car speed.

Note that X band radars are single frequency / channel (same frequency and frequency tolerance); any X band tuning fork will work with any X band radar. K band radars operate on one of 2 common (and overlapping frequencies offset by 25 MHz); tuning forks should be calibrated for a specific frequency (although the 2 K band frequencies are so close, and overlapping, in many cases a tuning fork works for both frequencies or at most off by 1 mph or kmh). Ka band radars can operate in a band of about 13 channels and require a tuning fork calibrated for the exact radar frequency.

Tuning forks should be checked periodically to insure resonance has not changed (due to a nick or bend from mishandling) or the fork mislabeled during previous lab calibration or other mix up. Every tuning fork should have documentation indicating fork **resonance** (and **frequency tolerance,** usually over **temperature**), radar **frequency** (X or K band, or exact frequency if Ka band radar) **and induced speed**. If the tuning fork does not have a calibration sticker and/or tractability, the accuracy is questionable.

Also see chapter 2.1—The Doppler Principle
Also see appendix C—Doppler Equations

Lab / Shop Calibration
Early microwave radars required a shop or laboratory periodically calibrate internal adjustments to insure accuracy and performance. Most later models

Donald S. Sawicki

(microwave and laser), due to cost and size reduction and greater speed of integrated circuits (IC's), are self-adjusting and/or self-testing and do not require periodic shop calibration.

Some agencies (state and/or local) require periodic (once or twice per year) additional checks and tests above and beyond everyday use testing. Some additional tests may or may not include, but not limited to measuring;

- entire speed range accuracy typically ± 1 mph stationary, ± 2 mph moving

- transmit frequency,
10.525 ± 0.025 GHz (X band)
24.125 ± 0.100 GHz (K band)
24.150 ± 0.100 GHz (K band)
33.4 to 36 ± 0.100 GHz (Ka band)

- transmitter power (milliwatts or dBm)

- ERP—effective radiated power (milliwatts or dBm)

- vertical and horizontal beamwidth
   typically 9° to 20° ± 1°

- receiver sensitivity (microvolts or -dBm)

- power supply limits (± volts)

- tuning fork resonance

- speed (mph, kmh, knots, etc.) versus Hertz

dBm—decibels (dB) above or below 1 milliwatt.

If additional testing is required records should document date, test conducted, and results (measurements). A calibration sticker (BY, DATE, DUE) should be attached to the radar indicating who calibrated the unit and when, and the date next calibration due. Any tuning forks used to check the radar (in everyday use) should also be checked and labeled for resonance at the time the radar is tested.

Traffic Radar Handbook
A Comprehensive Guide to Speed Measuring Systems

Chapter 5.3

Beam / Antenna Limitations

Outline
- Antenna Beam
- Displayed Target
- Dense Traffic
- Beam Pattern
- Moving Mode Clutter

Antenna Beam
At radar frequencies signal transmissions are of a line-of-sight (LOS) nature and do not bend or fringe around objects as do commercial AM and FM radio waves. Reflective objects mask and reflect radar signals, large non reflective obstructions can absorb and/or disperse radar signals.

Objects that mask (reflect) radar signals
—signs
—metal fences
—metal guardrails

Obstructions that absorb/disperse radar signals
—woods / forrest / jungle
—foliage
—hills / mountains
—stone / concrete structures (buildings, bridges, etc.)

The antenna beam is the major factor limiting traffic radars' ability to distinguish targets (no target range information available); any moving target in the beam may be detected by the radar. Detection is a function of target radar size and range from the radar. When multiple targets are in the beam, the operator must interpret (sometime guess) which target the radar is tracking.

Microwave traffic radar antenna beams vary from model to model about 9 to 25 degrees; the larger the antenna and/or the higher the frequency the more narrow the beam. Some Ka band across the road (photo radars) have a beamwidth of 5 degrees. The figure below graphs beamwidth spread in feet

Donald S. Sawicki

versus range for beamwidths of 1, 5, 10, 15, 20, 25, and 30 degrees. Most radar beams are not very selective except for extremely-short-range targets. The beam may easily cover several lanes of traffic at a relatively short range, increasing operator difficulty in discriminating targets.

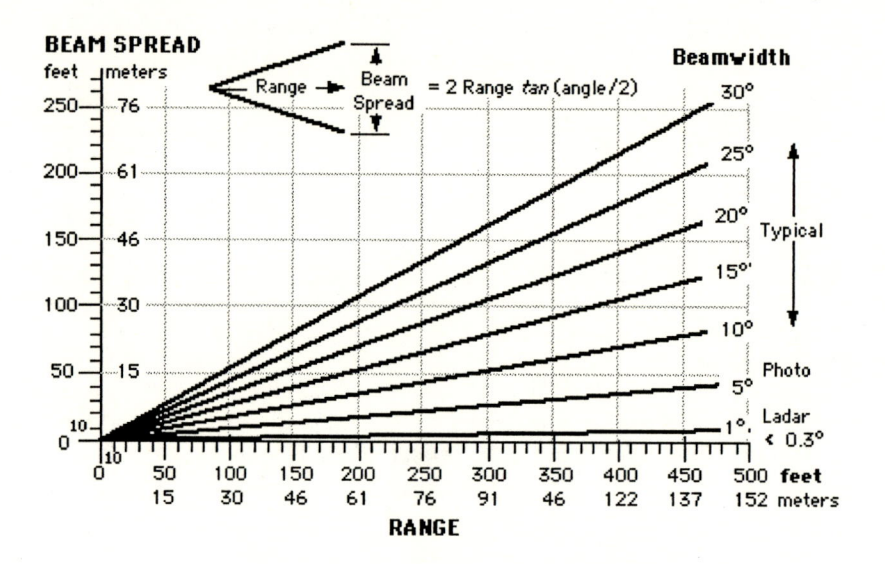

Figure 5.3-1—**Range vs Beamwidth**

Beam Spread = 2 Range tan (angle/2)

Displayed Target

Strongest Target Echo

Most radars display the speed of the target with the **strongest** echo (above a preset threshold—range control knob). The strongest target echo may be the closest target (in most cases), the biggest target, or something in-between the closest or biggest targets in the beam. A target's reflection varies with radar target size and target range from radar. A compact car close to the radar may have a smaller reflection than a tractor-trailer at a greater range (also see chapter 2.2—Radar Cross Section).

Fastest Target Tracking

Some radars allow the operator to track targets displayed based on strongest or **fastest** targets in the beam. Some of these radars have the option to display the speed of the **next fastest target** (after displaying the fastest target). The operator can select the next fastest target with the push of a button; each time the button is activated the next fastest target is displayed. Some radar models with same-lane capability (in same-lane mode) can toggle between the 2 fastest targets in the beam.

Strong echoes (from big close targets or other reflections) could reduce radar detection range by lowering receiver gain unbeknownst to the operator. If 3 or more targets are in the beam, the fastest target displayed may not be the expected target if the expected (fastest) target is at long range. The radar displays the fastest target in detection range.

Multi-Target Tracking

Some radar models can track and display speeds for **two different targets simultaneously**—the STRONGEST and FASTEST target in the beam (assuming the fastest target is not the strongest). Tracking 2 targets simultaneously could be a useful feature for the situation of a small car passing a large truck. Some of these radars allow the operator to select next fastest target (after displaying the fastest target).

Dense Traffic

Dense traffic makes it difficult, if not impossible, for the radar operator to distinguish which vehicle the radar is tracking at any given time. In fast moving dense traffic a different target could be displayed every radar update period— fractions of a second to seconds (see figure below). The problem of distinguishing targets is compounded if the radar tracks two targets (strongest and/or fastest and/or next faster…) in dense traffic.

Donald S. Sawicki

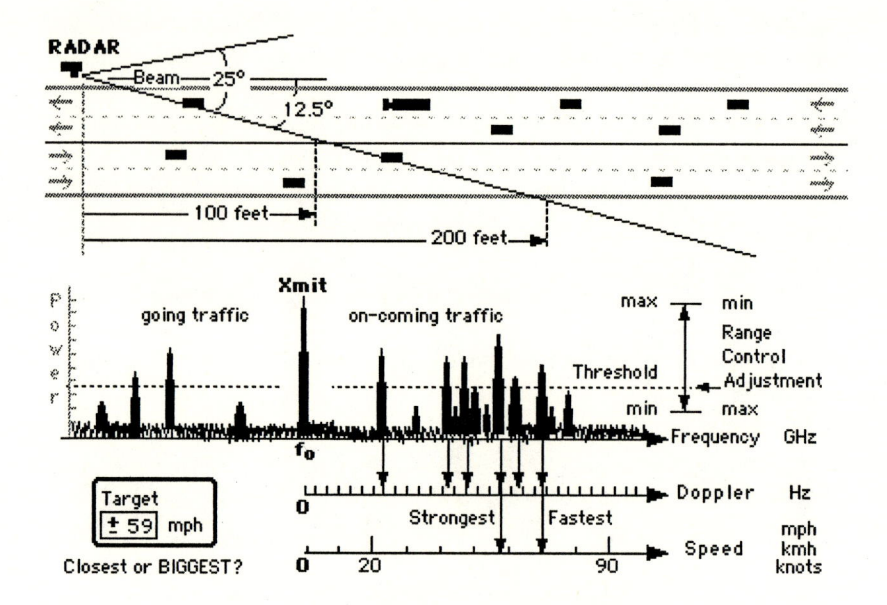

Figure 5.3-2—**Dense Traffic Scenario**
fo is radar transmit (Xmit) frequency

Beam Pattern

Radar has maximum target detection (range) when the target is in the beam center; at the beam edge (one-way beamwidth) detection range is 71% of detection in beam center (29 percent less range at beam edge). The beam edges are defined as the angle the one-way antenna gain is half of peak gain (-3 dB on a log scale); and the angle between the half power (-3 dB) beam edges is the beamwidth (by definition). Antenna beams are 3 dimensional and usually require two angles to describe the beam; one for the horizontal (azimuth) axis and one for the vertical (elevation) axis. Circular antennas have a symmetrical beam (horizontal beamwidth equals vertical beamwidth).

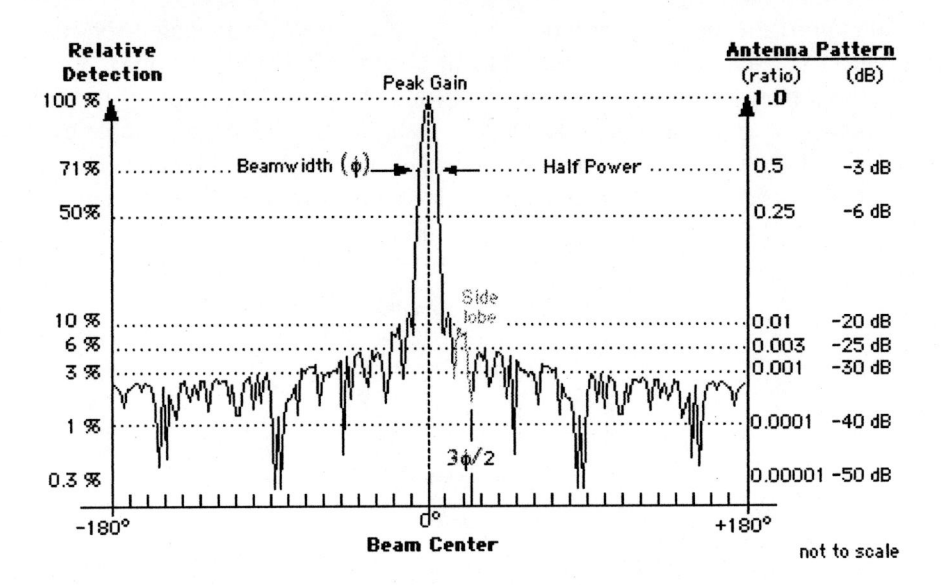

Traffic Radar Handbook
A Comprehensive Guide to Speed Measuring Systems

Figure 5.3-3—**Antenna Pattern vs Detection Range**

Antennas are not perfect; signals (although greatly reduced) can enter the antenna from almost any angle. For example the sidelobe closest to the maim beam is typically -20 dB (varies with antenna and frequency) of peak; -20 dB equates to a relative detection of 10 percent from that at beam center. This sidelobe typically occurs inside an angle (off beam center) less than 3/2 of beamwidth.

Note that the sidelobes of the antenna pattern are roughly the same shape (beamwidth) as the main lobe (the sidelobes do get a little wider as the angle gets farther off main beam). If the main lobe had a wider beamwidth the sidelobes would also be wider; conversely if the main lobe was more narrow the sidelobes would also be more narrow. The larger the antenna and the higher the frequency, the more narrow the beam.

Moving Mode Clutter

Stationary radar ground clutter, side or back lobes or main beam, is not Doppler shifted and therefore does not complete against legitimate targets. The limiting factor is receiver noise (KTBF)—also see appendix D—Radar Range Equation.

Moving mode radar ground clutter is Doppler shifted proportional to radar speed and the angle of the clutter echo. If clutter angle is 0 degrees (pointed in front of patrol car) main beam clutter (ground echo) corresponds to patrol car

Donald S. Sawicki

speed (V_p). If the angle is not 0 degrees a ground speed shadowing error may be introduced (see chapter 5.4—Operational Problems / Radar Shadowing).

Echo's from on-coming (and/or going) target's are in the clutter free part of the spectrum. Same-lane targets slower than radar patrol car complete with sidelobe clutter, targets faster than radar patrol car complete with backlobe clutter (unless the target speed is greater than twice patrol car speed). Also see chapter 2.1—Doppler Principle / Moving Radar Doppler.

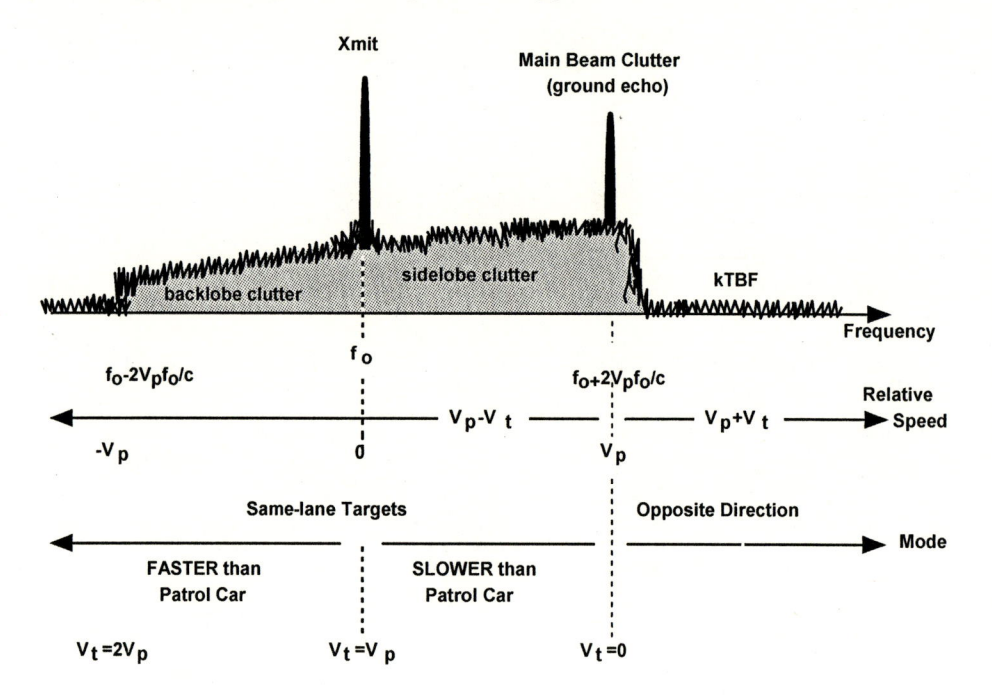

Figure 5.3-4—**Moving Mode Spectrum (Front Antenna)**

f_o = radar transmit frequency,
V_p = radar patrol car speed,
V_t = target speed,
c = speed of light,
KTBF = receiver noise floor.

Moving mode radar, both opposite direction target mode and same-lane mode, have difficulty measuring targets traveling close to zero (0) because main beam clutter (ground echo) spreads over a narrow band of frequencies centered at

Traffic Radar Handbook
A Comprehensive Guide to Speed Measuring Systems

patrol car speed. The frequency spread, and thus speeds covered, is a function of beamwidth, height above ground, and pointing angle.

Same-lane radars cannot measure targets traveling at or close to radar patrol car speed because the Doppler shift is at or near zero (0)—transmitter leakage interferes with signals not Doppler shifted in moving mode.

A rear facing antenna produces a spectrum that is the mirror image of a front facing antenna for moving mode radar.

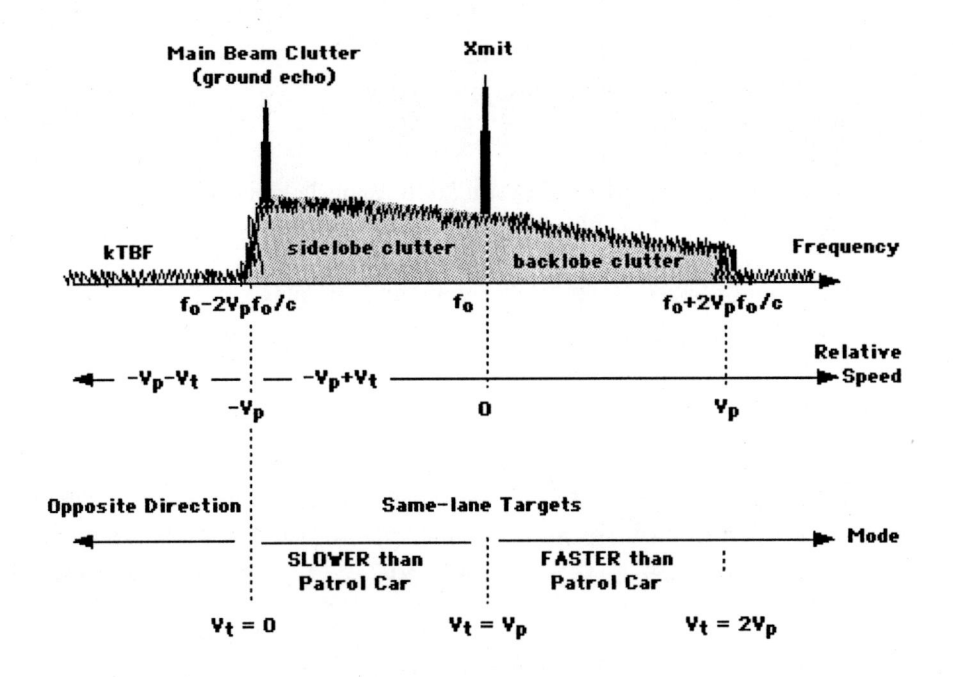

Figure 5.3-5—**Moving Mode Spectrum (Rear Antenna)**

Donald S. Sawicki

Chapter 5.4

Operational Problems

Outline
- Accuracy
- Cosine Effect Error
- Target Acceleration (Deceleration)
- Target Displayed (Target Discrimination)
- Target Harmonic
- Tracking Time
- Scanning Error/Hand-Held Modulation (hand-held models)
- Panning Error (multi-piece models)
- Patrol Speed Shadowing (moving mode radar)
- Batching or Speed Bumping (moving mode radar)
- Multi Mode Radar
- Auto-lock

Accuracy
Generally, under ideal conditions, most police microwave traffic radars are accurate to about **± 1 mph for stationary** operation and **± 2 mph for moving mode**. Moving mode radars also measure *patrol car speed* to an accuracy of about ± *1 mph*. Traffic radar should not be used to issue speeding tickets for alleged violations of 1 or 2 mph over the speed limit. Many police officers do not issue tickets (based on radar) unless the violation is at least 5 mph over the posted limit.

Cosine Effect Error
    see chapter 4.1—Cosine Effect Geometry
    see chapter 4.2—Cosine Effect on Measured Speed
    see chapter 4.3—Cosine Effect on Moving Radar

Target Acceleration (Deceleration)
Targets accelerating or decelerating, due to applying the gas or brakes, the cosine effect, and/or turning (see chapter 4.5—Cosine Effect on a Curve), more than about 1 mph (or 1 kmh) per radar integration cycle (sample time) may be difficult or impossible for the radar to track. A radar's ability to track a target depends on the target acceleration (or deceleration—negative acceleration)

component, radar integration cycle time (typically 0.1 to 2 seconds), and radar speed tolerance or accuracy (typically ±1 mph, or ±1 kmh, per radar integration cycle).

Maximum Target Acceleration or Deceleration
max target acceleration = speed accuracy / sample time

$$A_{tgt} = V_{tol} / t_o$$

A_{tgt} = target acceleration
V_{tol} = radar speed tolerance or accuracy (±1 mph typical)
t_o = radar integration period or sample time

The below table list maximum speed change per second (mph/sec, kmh/sec, or knots/sec) given radar sample time (in milliseconds or ms) and using an accuracy of ±1 mph, ±1 kmh, or ±1 knot). The change in speed (acceleration) is also equated to g's for mph, kmh, or knots. For example a radar with a sample time of 300 ms and an accuracy of ±1 mph cannot measure targets changing speed greater than about 3.3 mph/sec (about 0.15 g's); for an accuracy of ±1 kmh speed cannot change faster than 3.3 kmh/sec (about 0.09 g's).

RADAR ACCURACY mph or kmh or knots...	RADAR SAMPLE TIME	MAXIMUM SPEED CHANGE mph or kmh or knots... PER SECOND	g's mph	kmh	knots
+/- 1	2000 ms	0.5	0.02	0.01	0.03
"	1000 ms	1	0.05	0.03	0.05
"	800 ms	1.25	0.06	0.04	0.07
"	500 ms	2	0.09	0.06	0.10
"	360 ms	2.8	0.13	0.08	0.15
"	333 ms	3	0.14	0.09	0.16
"	300 ms	3.3	0.15	0.09	0.17
"	250 ms	4	0.18	0.11	0.21
"	200 ms	5	0.23	0.14	0.26
"	166 ms	6	0.27	0.17	0.32
"	143 ms	7	0.32	0.20	0.37
"	125 ms	8	0.36	0.23	0.42
"	111 ms	9	0.41	0.26	0.47

Donald S. Sawicki

| " | 100 ms | 10 | 0.46 | 0.28 | 0.52 |
| " | 50 ms | 20 | 0.91 | 0.57 | 1.05 |

Table 5.4-1—**Maximum Acceleration Given Sample Time**

The term g is an acceleration factor—based on the acceleration of a free falling body due to gravity. Note that 1 g, by International definition, is a change in speed **of 9.80665 meters per second every second**. Approximately **0.8 g's** (or greater) is very hard braking action. Adaptive Cruise Control (ACC) systems (vehicle radars) automatically brake when approaching other vehicles—maximum braking varies from 3 to 5 meters/sec^2 (7 to 11 mph/sec, or 10.8 to 18 kmh/sec), or about **0.3** to **0.5 g's**.

1 g = 9.8 m /sec^2 = 32.2 ft/ sec^2 = 21.9 mph/sec = 35.3 kmh/sec = 19 knots/sec

In general the longer the integration time (sample period) the greater the detection range, or stated another way the shorter the integration time the shorter the detection range.

Target Displayed (Target Discrimination)
 chapter 5.3—Beam/Antenna Limitations
 chapter 2.2—Radar Cross Section (target reflectivity)

Target Harmonic
Strong target echoes (**large targets close to radar**) have the potential to saturate (overpower) the radar receiver (low noise amplifier and/or mixer) producing harmonics of the target echo (see chapter 5.5—Harmonic Interference). If the radar locks onto a harmonic, the displayed target speed could be **double** or **triple** true speed!

Multiple large targets close to the radar can produce harmonics with enough power to produce other harmonics (that may produce other harmonics). With so many harmonic signals the radar could produce almost any speed reading depending on harmonic power. Moving mode radar has the potential to generate even more harmonics because the ground echo is frequency shifted (another frequency and potential harmonic generator).

Tracking Time
The longer a radar tracks a target, the better the chance for an accurate measurement and the lower the probability of interference from another target, moving object, or transmitter (chapter 5.5—Interference). Faster targets are

inherently easier for the radar to track and measure, but generally have shorter times for tracking (max and min radar range constraints). Because most radars do not have any sort of written record (such as a strip chart), there is no way to determine exactly how long the radar was really tracking any particular target, especially in a target rich environment—dense traffic (figure 5.3-2—Dense Traffic Scenario). Pulsed radars, by design, transmit just long enough (fractions of a second to a second or so) to process and display a target speed; this may make it difficult to know for sure which target the radar was tracking if multiple targets are in the beam.

Scanning Error / Hand-Held Modulation (hand-held models)

A way to unintentionally (or intentionally) produce an error on the order of a few mph is to quickly move, or shake, (modulation) the radar (antenna) while tracking a target. The error introduced by radar motion is referred to as scanning error or hand-held modulation and can be a problem with single piece models, or models with a hand-held antenna. A false target can also be produced by pointing or scanning the radar antenna at or near the dash board allowing the heater / air-conditioning fan (if moving) to register as a target (see chapter 5.5—Interference / Mechanical Interference).

Target speed error is a function of radar sample time (integration period), the rate or speed the radar antenna moves, and the direction of antenna movement with respect to target. The **shorter** the radar sample time, the **faster** the antenna moves, and the more any antenna movement is **directly toward** (or **away**) from target, the greater the error. Antenna movement **directly toward or away** from target translates directly into speed error, speed antenna moves is speed error (error high when antenna moves toward target, low when moves away).

RADAR SAMPLE TIME	MOVEMENT DURING SAMPLE TIME FOR 1 mph ERROR		MOVEMENT DURING SAMPLE TIME FOR 1 kmh ERROR	
ms	in	cm	in	cm
1000	17.6	44.7	10.9	27.8
900	15.8	40.2	9.8	25.0
800	14.1	35.8	8.7	22.2
700	12.3	31.3	7.7	19.4
600	10.6	26.8	6.6	16.7
500	8.8	22.4	5.5	13.9
400	7.0	17.9	4.4	11.1
360	6.3	16.1	3.9	10.0
333	5.9	14.9	3.9	9.3
300	5.3	13.4	3.3	8.3
250	4.4	11.2	2.7	6.9
200	3.5	8.9	2.2	5.6
166	2.9	7.4	1.8	4.6
143	2.5	6.4	1.6	4.0
125	2.2	5.6	1.4	3.5
111	2.0	5.0	1.2	3.1
100	1.8	4.5	1.1	2.8
50	0.9	2.2	0.5	1.4

ms - milliseconds, in—inches, cm—centimeters

Table 5.4-2—**Antenna Movement Error**

For example a radar with a sample time of 100 milliseconds that has a quick forward motion of 1.8 inches during radar sample time (0.1 seconds) produces an error 1 mph high (1 mph low if radar moves backward). To produce a 2 mph error radar movement would be 2 times 1.8 inches or 3.6 inches.

Conversions
1 mph = 17.6 in/sec = 44.704 cm/sec
1 kmh = 10.9 in/sec = 27.788 cm/sec
1 knot = 20.3 in/sec = 51.444 cm/sec

The faster the radar sample time the greater the chance of introducing a scanning error. Error magnitudes are on the order of a few mph (km/h, or

NM/h); faster radar motion will tend to drop legitimate targets to track ground movement (resulting from radar movement). For very slow speeds traffic radar cannot measure ground speed because the Doppler shift is so small and overlaps the transmitter signal. Also, the shorter the sample period the more spectrum the main signal covers and the greater the minimum speed required to get a measurement.

Panning Error (multi-piece models)
Two-piece radar units will produce false readings if the radar antenna is scanned past the electronics (display). Scanning the antenna at different rates or angles can also produce different false readings (panning).
Patrol Speed Shadowing (moving mode radar)

Shadowing from Stationary Objects

Figure 5.4-1—**Shadowing**

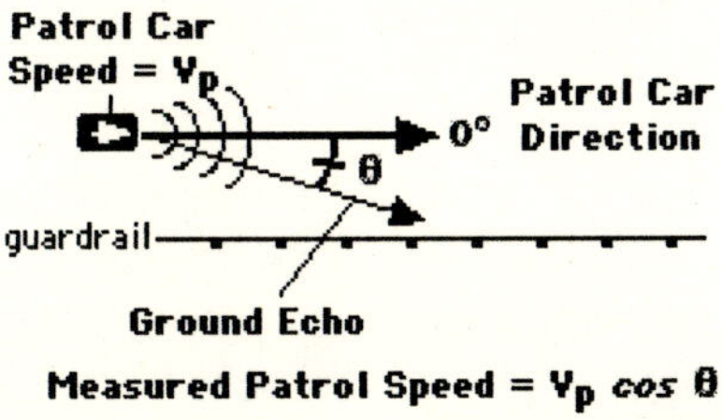

Moving mode radar derives target speed from patrol car speed (computed from ground echoes). For the radar to measure exact patrol car speed the ground echo should come from the front of the patrol car (0 degrees). Ground echoes from either side of the patrol car (shadowing) measure patrol car speed LOW due to the Cosine Effect (see chapter 4.1). Note that it does not matter which side (right or left) the ground echo returns from, only the magnitude (angle); the greater the angle the lower the measured patrol car speed. Measured patrol car speed equals actual speed times the cosine of the angle of the ground echo (patrol car direction is 0 degrees). When the angle is 0 degrees, radar measured patrol car speed equals actual speed (no error).

Patrol Car Speed = V_p
Patrol Car Direction 0°
guardrail
Ground Echo
Measured Patrol Speed = $V_p \cos \theta$

Measured Patrol Speed = Vp *cos* (theta)
V_p = patrol car speed, theta = angle of ground echo,
if V_{err} = speed error (patrol speed low by Verr)
theta = *arccos* (1 - V_{err} / V_p)

With same-lane radar a low patrol speed reading translates into a low target speed reading. For small radar-target Cosine Effect angles target speed measures low by the error in measured patrol speed. If radar patrol speed measures low by -**Verr** mph, target speed computes low by -**Verr** mph. If the Cosine Error is significant the measured target speed is even lower. Also see chapter 2.1—Moving Radar Doppler.

However with **opposite direction (targets)** a low patrol speed reading translates into a high target speed reading for small radar-target Cosine Effect angles (see chapter 4.3). If measured patrol car speed has a -**Verr** mph error, measured target speed has an error of +**Verr** mph. For example if the radar measures patrol car speed low by 3 mph, computed target speed is high by 3 mph. This formula applies to moving radar (opposite direction targets) with either forward or rear facing antennas.

Opposite Direction Traffic
Negligible Radar-Target Cosine Effect Angle
V_p = patrol speed, V_t = target speed, V_{err} = speed error;
patrol speed low by V_{err}.

Measured Patrol Speed = V_p - V_{err}

Measured Target Speed = V_t + V_{err}

Radars identify ground echoes as the slowest strongest signal (it is most of the time). The ground echo cosine angle (theta) is a function of the radar antenna alignment and beamwidth. More reflective terrain in only part of the beam could change the angle (off beam center) of the ground return (shadowing) which would change the measured patrol car speed slightly. If terrain is reflective enough and the radar is close enough, the strongest signal could even creep into the receiver from out of the main beam (through a sidelobe)—causing significant errors due to a large cosine angle. Also see chapter 5.3—Beam/Antenna Limitations / Moving Mode Clutter.

Large extended objects can cause patrol speed shadowing for an extended period, and include;

- **guardrails**
- **bridge trusses**
- **hills / mountains**
- **plowed snow pile**
- **multiple parked vehicles**
- **construction zones** (with lots of signs)

Passing other large objects can have a momentary affect on measured patrol speed, and include;

- overpasses
- billboards
- road signs
- overhead signs
- parked vehicles
- ice patches, snow piles

Shadowing from Moving Vehicles

Moving objects can cause patrol speed shadowing. If a radar mistakes a slow (slower than patrol car speed) vehicle (same direction as patrol car) for the ground echo, opposite direction targets measure high (by the speed of the false ground target). The radar measures patrol car speed low (by the false ground target speed).

Rear Antenna
Target Speed = V_t
Patrol Car Speed = V_p
False Ground (Target) Speed = V_{t2}

$V_p > V_{t2}$

Forward Antenna
Patrol Car Speed = V_p
Target Speed = V_t
False Ground (Target) Speed = V_{t2}

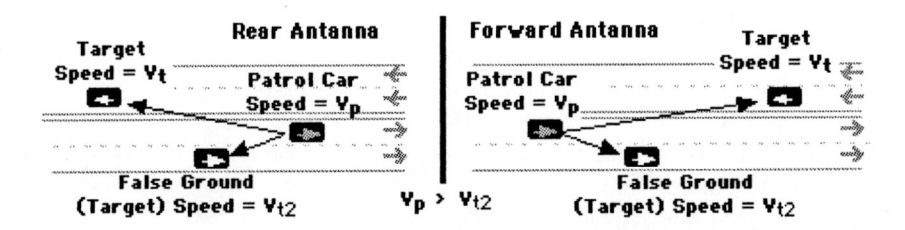

Figure 5.4-2—**Moving False Ground Echo**

Opposite Direction Traffic
Negligible Radar-Target (V_t) Cosine Effect Angle
Radar Mistakes Moving Target (V_{t2}) for Ground Echo
V_p = patrol speed, V_t = target speed, V_{t2} = false ground target speed.

$$\text{Measured Patrol Speed} = V_p - V_{t2}$$

$$\text{Measured Target Speed} = V_t + V_{t2}$$

For example if a 35 mph radar patrol car mistakes a 20 mph vehicle for a ground echo, the radar calculates patrol car speed 20 mph low (35-20=15). Opposite direction targets measure high by 20 mph, the error in patrol car speed which is false (ground) target speed. Note that vehicles passing (faster than) a patrol car have a negative Doppler shift (negative patrol car speed) and should not be mistaken for a ground return.

If a radar mistakes a slow (slower than patrol car speed) vehicle (same direction as patrol car) for the ground echo, **same-lane** (direction) targets measure LOW (measured target speed is the difference in speed between the two targets). The radar measures patrol car speed low (by the false ground target speed). False (ground) target speed must be less than patrol car speed and target speed.

To eliminate shadowing from stationary or moving objects some moving radars measure patrol car speed using the patrol car **speedometer cable** instead of radar ground return. With this configuration some radars also automatically switch between moving and stationary mode without operator input. This system requires the operator calibrate the patrol car speedometer to radar by adjusting radar measured speedometer (cable) to radar ground echo while moving. However if the speedometer (cable) is not linear over the speed range additional errors are probable. Also spinning or locked wheels as well as acceleration or deceleration will introduce batching errors (see below).

Batching or Speed Bumping (moving mode radar)

Moving radar false or inaccurate target display readings are possible if the patrol car suddenly changes speed (accelerates or decelerates). Radar-measured target speed and patrol car speed (from ground echoes) are not updated simultaneously; if patrol car speed changes suddenly the radar may still be using outdated patrol speed data leading to a momentary measured speed error (batching or speed bumping).

Multi-mode Radar

Some radars measure on-coming and/or going traffic while the radar is either stationary or moving. Some moving radars can also measure targets traveling in the same direction (same-lane) as the patrol car. Some systems have two antennas, one mounted in the front (pointed forward) and one mounted in the back (pointed aft). Antennas can be mounted either inside or outside the patrol car. Some systems can use either the front or the rear antenna alone, or both antennas simultaneously.

Stationary radars set to measure both on-coming and going traffic require the operator to observe targets from 2 directions (on-coming and going). Some radars (single antenna operation) indicate displayed target speed direction (on-coming or going). In a two antenna system (forward and rear looking) targets from 1 to 4 directions are possible.

Traffic Radar Handbook
A Comprehensive Guide to Speed Measuring Systems

FORWARD ANTENNA
on-coming (opposite direction) traffic
going (same direction) traffic

REAR ANTENNA
on-coming (same direction) traffic
going traffic (opposite direction)

Moving radars with same-lane capability and opposite direction tracking, and a single antenna (forward OR rear looking) have 1 to 2 possible target directions (some systems indicate displayed speed direction—same direction (same-lane) or opposite direction). Radars with dual antennas (front AND rear) have 1 to 4 possible target directions.

FORWARD ANTENNA
opposite direction (on-coming) traffic
same-lane traffic—front of patrol car

REAR ANTENNA
opposite direction (going) traffic
same lane traffic—rear of patrol car

The ability to measure and display targets from 2 or more directions may make it difficult for an operator to know which target a radar is tracking when multiple targets are in the beam(s).

Auto-Lock

Auto-lock freezes (holds) the target display automatically when a preset target speed has been met or exceeded. Unattended photo and/or video radars depend on an auto-lock feature; some older manual radars have this feature (later models do not). With an Auto-lock no one has to have observed the violation, the process is automatic. Some states and courts do not allow the use of this feature even if the radar has one.

The main disadvantage of an auto-lock is that it does not require an operator (or anyone) to see the violation. Another serious disadvantage is the possibility of the radar automatically locking onto a momentary stray signal (not that uncommon). In moving mode radar auto-lock does not allow time for the operator to compare radar measured patrol car speed to patrol car speedometer reading.

Donald S. Sawicki

Chapter 5.5

Interference

Outline
- Radio Frequency (RF) Interference
- Receiver Saturation
- Harmonic Interference
- Mechanical Interference
- Natural Interference
- Unintentional Forms of Radiation

Radio Frequency (RF) Interference
Interference may occur when a receiver is close to another transmitter. The closer the transmit frequency to the intended receive frequency, and the closer the transmit and receive antenna, and the stronger the transmit signal, the greater the chance of interference. Transmitters have frequency components off center frequency; these off-frequency skirts are relatively close to the center frequency and reduced in power. Transmit signal power, modulation, and bandwidth determine the shape and extent of the signal spectrum. If the traffic radar is close enough and the interfering transmit signal is strong enough a transmitter not exactly on radar frequency may cause interference.

Typical High Power Transmitter Sources

- airports
- hospitals
- police / fire stations
- cable TV companies
- cellular / telephone companies
- weather radars
- microwave relay towers
- radio towers

Strong close signals can effect a traffic radar receiver, analog circuits, and/or digital circuits—causing malfunctions, blanking (out), or false readings. Radio transmitters in and near the radar patrol car that can cause interference include but not limited to;

Traffic Radar Handbook
A Comprehensive Guide to Speed Measuring Systems

- **100 watt** police radios (VHF/UHF)
- **2 watt** police walkie-talkies (VHF/UHF)
- **High Frequency** police radios (HF)
- **CB** transmitters (HF)
- **Cellular Phones** (UHF)

Mutual Interference—Two (or more) traffic radars transmitting at on near the same frequency operating in close proximity may interfere with each other. One (or more) radar may interfere and/or JAM the other(s). The degree of interference depends on radiated power, antenna pointing angles, exact frequencies, and distance between radars.

Some **FIELD DISTURBANCE SENSORS** such as

Intrusion Sensors:	burglar alarms
Automatic Door Openers:	grocery / store, railroad train, truck / automobile entrances.
Obstruction Detectors:	—used by (mounted to) fork lifts moving farm equipment railroad locomotives / rolling stock, specialized rail cars
Object Detectors:	production line counting / sensing

transmit at X or K band traffic radar frequencies (see figures below). Field disturbance sensors (both bands) are allowed by the FCC to radiate 2.5 V/m at 3 m (9 ft, 10 in) or about 1.9 watts ERP; greater than a typical X or K band traffic radar ERP of 100 mW (chapter 5.1—NHTSA 1980 Test). Field disturbance sensor maximum limits have about 19 times greater (over +12 dB) ERP, and have **4 times the RANGE** (to produce the same or greater power density), then a typical X or K band traffic radar. Also see chapter 8.3—Power Density.

FCC Rules and Regulations, Part 15, Subpart C: Intentional Radiators

FIELD DISTURBANCE SENSOR LIMITS (Section 15.245, 1996 Aug 16)
10.525 GHz ± 25 MHz, 2.5 V/m at 3 m

Donald S. Sawicki

24.125 GHz ± 50 MHz, 2.5 V/m at 3 m

LOW POWER FIELD DISTURBANCE LIMITS (Section 15.249, 1990 Jun 20)
24.125 GHz ± 125 MHz, 250 mV/m at 3 m

Field Disturbance Sensors are CW (continuous wave), these devices may interfere with a traffic radar by

• reducing radar detection range
(automatic gain control adjust out interference and detection range),

• activating radar interference detected circuits
(some models stop processing targets if interference detected),

• masking a legitimate target
(masking multiple Doppler filters),

• producing false speed readings
(if interfering signal falls in a single Doppler filter).

The first two items depend on Sensor signal strength, the last two also depend on frequency and bandwidth. Generally the stronger and/or closer the interfering signal is to the radar, the greater the degree of any effects.

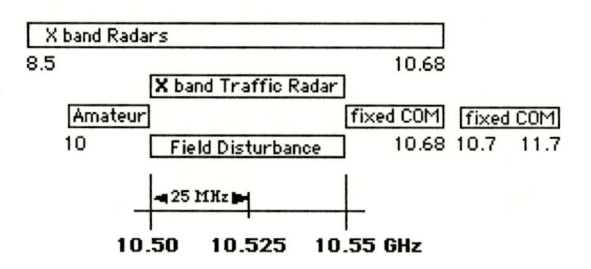

Figure 5.5-1—**X band field disturbance**

Ground and airborne X band radars operate between 8.5 - 10.68 GHz. Weather and navigation radars usually operate between 9.3 - 9.5 GHz; precision approach radar operates between 9 - 9.2 GHz. Fixed COM indicates stationary communications (as opposed to mobile).

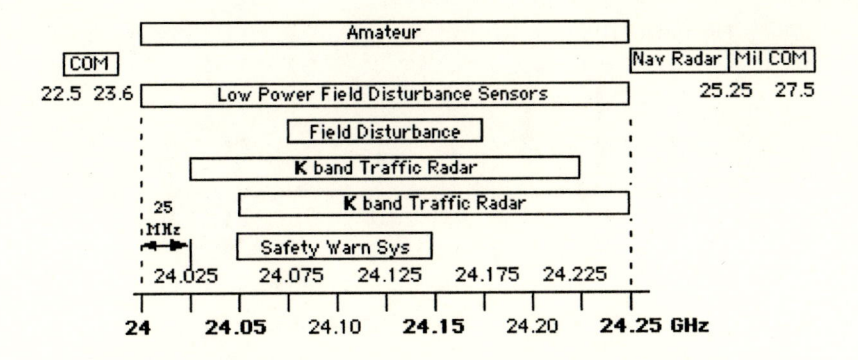

Traffic Radar Handbook
A Comprehensive Guide to Speed Measuring Systems

Figure 5.5-2—**K band field disturbance**

Low power (250 mV/m at 3 m) field disturbance sensors include unlicensed / unattended radars, and some sports radars. COM indicates communication, mil indicates military use, and Nav indicates navigation (radar).

Safety Warning Systems transmit at 24.1 GHz ±50 MHz and are intended for reception by some radar detectors and special receivers; Safety Alert systems transmit at 3 frequencies (24.07, 24.11, 24.19 GHz). Also see chapter 7.1—Counter Measures / Radar Detectors.

Also see appendix A, figure A-1—Select Ka Band Radars.

Receiver Saturation

Receiver saturation occurs when a signal is greater in power than a receiver was designed to process. The resultant cross-modulation produces spurious signals and harmonics (of interfering signal, transmitter, local oscillator, and/or any target echoes) that can mask real signals or confuse tracking circuits. Results include greatly reduced radar sensitivity (detection range), or tracking a spurious signal. Any receiver operating close to a high powered transmitter risks saturation.

Harmonic Interference

Another way nearby transmissions may interfere with traffic radar is from a harmonic. Radio and microwaves have harmonics, whole multiples of the fundamental frequency (see figure below). For example, a fundamental signal at 2 GHz has a second harmonic at 4 GHz and a third harmonic at 6 GHz. Generally, the higher the harmonic the lower the power, but harmonics may still have sufficient power to cause interference.

Donald S. Sawicki

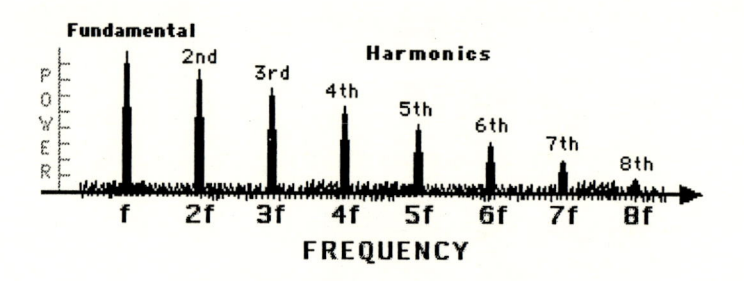

Figure 5.5-3—**Harmonics**

The tables below illustrate harmonic interference for X, K, and Ka band radars. The tables show the fundamental frequency (or band for Ka radar) and associated harmonic that could interfere with traffic radar. Designated uses of the interfering signal are also shown. For example, an S band radar at 3.508 GHz has a third harmonic at the X band traffic radar frequency (3.508 x 3 = 10.524).

ABBREVIATIONS: Com—communications, DBS—Direct Broadcast Satellite, LAN—Local Area Network, LOS—Line-of-Sight.

Band	Frequency (GHz)	Interfering Harmonic	Source(s)
X	10.525 ± 0.025	1st	Traffic Radars, Field Disturbance Sensors
C	5.2625 ± 0.0125	2nd	LANs, C Band Radar, Landing Systems
S	3.508 ± 0.008	3rd	Auto Vehicle ID, Amateur, S Band Radars
S	2.631 ± 0.006	4th	Wireless Cable, DBS, Instructional TV
S	2.105 ± 0.005	5th	LOS Com
L	1.754 ± 0.004	6th	Military Com, Troposcatter Telemetry
L	1.5036 ± 0.0036	7th	Telemetry, Navigation, Com

Table 5.5-1—**Harmonic Interference and X Band Radar**

Band	Frequency (GHz)	Interfering Harmonic	Source(s)
K	24.150 ± 0.1 24.125 ± 0.1	1st	Traffic Radars, Field Disturbance Sensors
Ku	12.075 ± 0.05 12.063 ± 0.05	2nd	LOS Com
X	8.050 ± 0.03 8.042 ± 0.03	3rd	Military Com
C	6.038 ± 0.025 6.031 ± 0.025	4th	LOS Com, Satellite Uplink
C	4.830 ± 0.02 4.825 ± 0.02	5th	Moblie Radio, Military Com
C	4.025 ± 0.016 4.020 ± 0.016	6th	LOS Com
S	3.450 ± 0.014 3.446 ± 0.014	7th	Auto Vehicle ID, Amateur, S Band Radars

Table 5.5-2—**Harmonic Interference and K Band Radar**

K band traffic radars operate at;
24.150 GHz ±100 MHz (24.050 - 24.250 GHz), or
24.125 GHz ±100 MHz (24.025 - 24.225 GHz).

Band	Frequency (GHz)	Interfering Harmonic	Source(s)
Ka	33.4 - 36.0	1st	Traffic Radars
Ku	16.7 - 18.0	2nd	Ku Band Radar, LOS Com, DBS Uplink
X	11.1 - 12.0	3rd	Military Com
X	8.35 - 9.00	4th	LOS Com, Satellite Uplink
C	6.68 - 7.20	5th	Mobile Radio, Military Com
C	5.567 - 6.0	6th	Microwave Ovens, Amateur, LOS Com, Satellite Uplink
S / C	4.77 - 5.14	7th	Fixed and Mobile Com

Donald S. Sawicki

Table 5.5-3—Harmonic Interference and Ka Band Radar

Com—communications
LOS—Line-of-Sight
DBS—Direct Broadcast Satellite

Airports, military installations, telephone communication centers, cable companies, hospitals, fixed and mobile microwave transmitters, microwave relay towers, satellite up-links, aircraft, ships, and amateur radio transmitters are just a few possible sources of interference.

Mechanical Interference
Traffic radars are capable of measuring any reflective moving object. A phenomenon well known among many radar operators is that pointing the radar antenna at the police car heater/air conditioning fan (while rotating) can produce false readings. The radar picks up the motion of the fan blades; by slightly changing the angle of the radar antenna a different speed reading can be obtained. Objects that can produce readings include but not limited to;

- **police car heater/air conditioning fan**
- **moving signs** (highway billboards)
- **building/house air conditioning fans**
- **wind gauges**

Natural Interference
Temperature affects the sensitivity of all receivers: the lower the temperature, the better the sensitivity (longer range). The sun affects background noise; as the sun heats objects, the thermal excitation of atoms in conducting materials generates noise that can obscure signals and reduce a receiver's performance. Lightning can also interfere with a receiver's performance.

Below 100 MHz atmospheric attenuation is negligible, above 5 GHz it becomes significant, and above 20 GHz it is profound. The higher the frequency the more signal energy the atmosphere absorbs due to the resonant frequencies of oxygen (O_2) and water (H_2O) molecules in the air. The higher the concentration of oxygen (higher altitudes have less loss) and water (rain, snow, sleet, ice, fog, and/or humidity), the more energy absorbed. K band radar is especially vulnerable to water vapor (humidity). The atmosphere has a significant affect on K and Ka band radars, a much lesser affect on X band radars.

The figure illustrates signal loss due to rain,, water vapor or humidity (H_2O), oxygen (O_2), and fog—total loss is the product of all of the above. The more

Traffic Radar Handbook
A Comprehensive Guide to Speed Measuring Systems

moisture the more the water vapor curve shifts upward (greater losses). Note: water vapor causes peak losses at 22.24 and 184 GHz; oxygen causes peak losses at 60 and 118 GHz. Transmission windows (between O_2 and H_2O resonant losses) exist around 35 GHz in the Ka band, and in the millimeter (mm) band around 94, 140, and 220 GHz. Note the 94, 140, and 220 GHz transmission windows have greater loss than the 22.24 GHz H_2O absorption band.

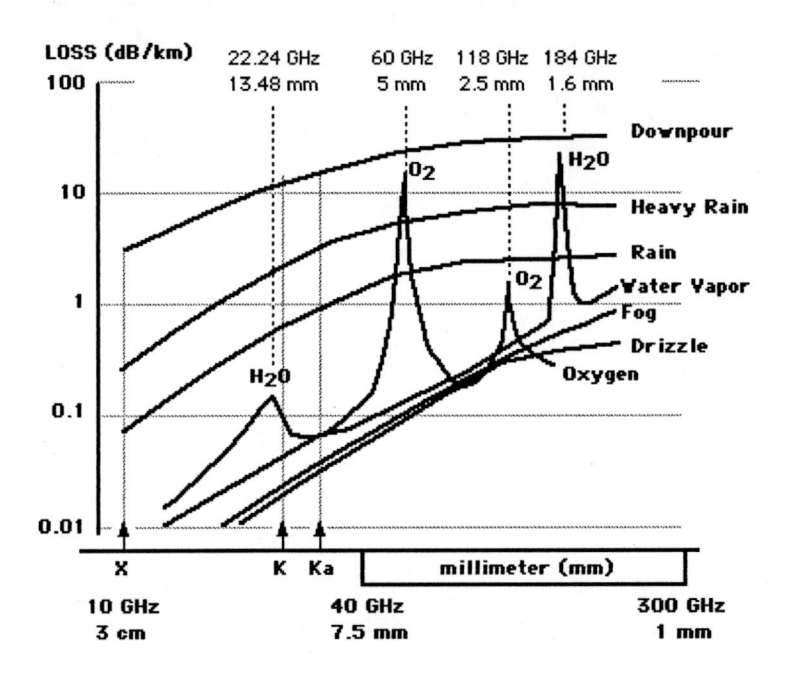

Figure 5.5-4—**Atmospheric Losses**

CONDITIONS: One-way attenuation (dB/km = dB/0.62 mi = dB/3281 ft = dB/1094 yds) in clear atmosphere,1% water vapor molecules for 7.5 g/m^3), 76 cm pressure. Downpour—6 in/hr = 150 mm/hr, Heavy Rain—1 in/hr = 25 mm/hr, Rain—0.2 in/hr = 5 mm/hr, Drizzle—0.01 in/hr = 0.25 mm/hr, Fog—0.1 g/m^3 (50 m = 164 ft visibility).

Unintentional Forms of Radiation

Other sources of interference include **ignition, alternators, spark plugs**, and **wiper motors**, those on the radar vehicle and on vehicles close to the radar. The ignition draws a lot of current (for a short time); the more current, the higher the electromagnetic field levels. Alternators are electromagnetic devices by design;

fields vary with engine rpm (revolutions per minute). Spark plugs generate a great deal of noise by generating high powered pulses that fire relatively fast. Wiper motors generate strong electromagnetic fields that change with time (wiper rate).

The switching on or off of **fluorescent, mercury, sodium vapor** (common for street lighting) and **neon / argon** lights (many designed to switch on and off to attract attention) can also trigger false radar readings. During switching time the changing current creates a changing electromagnetic field with lots of frequency components; the greater the current the stronger the field.

High voltage power lines and/or **transformers** are a possible source of interference if the radar is close enough and/or antenna pointed at or near direction of high voltage source. This type of interference usually causes a buzzing or humming of the radar audio Doppler.

Chapter 6.1

Introduction to Laser Radar

Laser radars, also referred to as ladar or lidar, use pulsed laser light instead of continuous microwaves to sense a target. Lasers are extremely pure (coherent) lightwaves, similar to only one (pure) color of light. Note that white light consists of multiple wavelengths (colors) with random (non-coherent) phases. Theodore Maiman of Hughes Aircraft Company (California) built the first working laser using a ruby rod pumped by a flash lamp in May of 1960. Traffic ladars use solid-state diodes to generate laser light.

LASER - **L**ight **A**mplification by **S**timulated **E**missions of **R**adiation
RADAR - **RA**dio **D**etection **A**nd **R**anging
LADAR - **LA**ser **D**etection **A**nd **R**anging
LIDAR - **LI**ght **D**etection **A**nd **R**anging

Laser radars transmit pulsed laser light to measure target range. The time it takes for a laser light pulse to travel (at the speed of light) from the ladar to the target and back is used to compute the distance from the ladar to target and back (distance pulse travels = speed of light x time). Target range from ladar is half of this distance (Range = 0.5 x speed of light x time). The change in target range over time (1/3 second typical) equals target velocity. Laser radar must transmit a minimum of 2 pulses to get at least 2 range measurements at 2 different times to compute speed. In reality laser radars transmit tens to hundreds of pulses per second.

Apertures for ladars are optical focusing devices (lenses, prisms, and/or mirrors) used to collimate laser energy into a narrow beam. Some models use the same aperture for transmit and receive; some use separate apertures (one for transmit and one for receive). The LTI 20-20 laser radar has two separate apertures; one on top of the other (top aperture transmits). The LR 90-235/p (Europe) has two separate apertures: side-by-side.

Ladars use a semiconductor diode (typically 3 diodes) to generate laser light. Most traffic ladars emit laser light around 904 nm wavelength. Other wavelengths are possible; for example aluminum gallium arsenide (AlGaAs) diodes emit light at a wavelength of 850 nm (some fiber optics use this wavelength). Gallium arsenide (GaAs), classified as an injection laser, emits light between 880 nm to 900 nm between the temperatures of -20 and 140 degrees Fahrenheit. Other wavelengths are possible using other materials or alloys.

Government Regulations

The FCC regulates radiated emissions from high speed circuits such as processing circuits inside a ladar, but not light frequencies. The Federal Drug Administration (**FDA**) Center for Devices and Radiological Health (**CDRH**) regulates laser products sold in the United States. Traffic ladars are class 1 devices (by American National Standards Institute definition) and considered eye-safe based on current medical knowledge—even so it is probability a good idea NOT to stare at a ladar aperture while transmitting, especially at close range.

Traffic Radar Handbook
A Comprehensive Guide to Speed Measuring Systems

Chapter 6.2

Light
Spectrum

Light is generally measured in wavelength (distance per cycle) instead of frequency (cycles per second). Common wavelength dimensions (metric) include angstroms (ang or Å), nanometer (nm), and micrometers (microns or µm).

$$1 \text{ ang} = 10^{-10} \text{ meters}$$
$$1 \text{ nm} = 10^{-9} \text{ meters}$$
$$1 \text{ µm} = 10^{-6} \text{ meters}$$

	——meters——	——inches——
1 angstrom (ang)	= 0.000 000 0001	= 0.000 000 0039
1 nanometer (nm)	= 0.000 000 001	= 0.000 000 0393
1 micron (1 micrometer)	= 0.000 001	= 0.000 039 3701

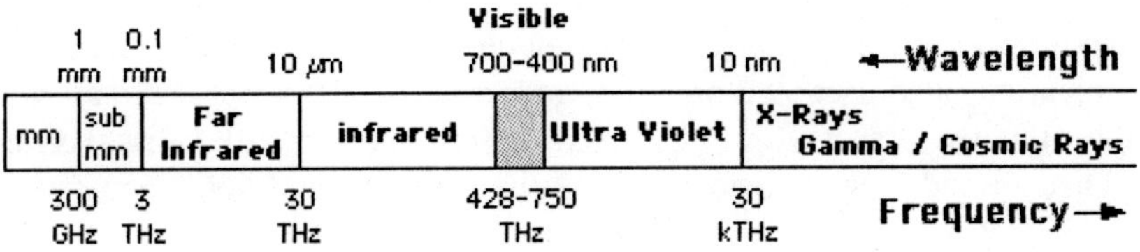

mm	sub mm	Far Infrared	infrared	Ultra Violet	X-Rays Gamma / Cosmic Rays
1 mm	0.1 mm	10 µm	Visible 700-400 nm	10 nm	←Wavelength
300 GHz	3 THz	30 THz	428-750 THz	30 kTHz	Frequency→

Figure 6.2-1—**Upper Spectrum**

Ladars operate around 904 nm wavelength, in the near infrared (IR) region—invisible to the human eye. The infrared region is between millimeter (mm) waves and visible light. We sense infrared energy as heat. Visible light is between about 700 nm (red) and 400 nm (violet). The figure below shows graphically the IR and visible light spectrum and lists several uses.

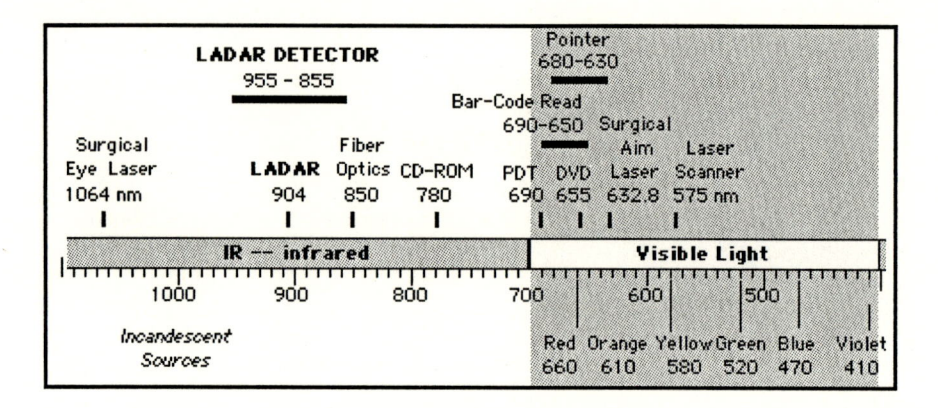

Figure 6.2-2—**Light Wavelengths**

Early **laser radar (ladar) detectors** responded to wavelengths between 400 - 1100 nanometers (nm); later models narrowed the spectrum to about 905 nm ± 50 nm. **Fiber optic** communications systems use the wavelengths of 850 nm (shown in the figure), 1310 nm, and 1550 nm. Some **surgical eye lasers** operate at 1064 nm, near the incandescent peak (around 1000 nm). Some **surgical aiming lasers** operate in the visible spectrum at 632.8 nm (lower orange part of the spectrum). Lasers that read **CD-R/RW** (Compact Discs—Read Only / Read and Write) operate at 780 nm; one **free-space optics** system (over the air optical communication) also uses 780 nm (fog, and rain to a lesser degree, may interfere with the signal). Lasers for **DVD-R/RW** (Digital Versatile Discs or Digital Video Disc - Read Only / Read and Write) operate around 660 - 655 nm (in the visible red part of the spectrum). Laser **pointers** operate between 630-680 nm. Many laser **bar-code readers** operate around 675 nm ± 5 nm. One type of **PDT** (PhotoDynamic Therapy) for treating cancer uses laser light at 690 nm to activate injected drugs. Many **Test and Measurement** lasers operate at 635 nm.

Chapter 6.3

Laser Radar Operation

Outline
- Test Before / After Use
- Beamwidth
- Operation
- Detection Range

Laser radars (ladars) operate from a stationary position (no moving mode) and measure velocity (approaching and/or receding traffic); some models have the option to display target range also. Laser radars also can measure the range of stationary targets.

Test Before / After Use
Ladars measure small range changes in short periods of time (see appendix F—Ladar Technical Details)—this imposes tight tolerances on ladar internal timing circuits. The units should be calibrated (by a calibration lab) on a regular basis. To help insure a ladar is functioning properly, before and/or after use (measurement and/or shift) the operator should;

- run ladar **self-test**
- check **beam to sight alignment** (horizontal and vertical axis)
  A telephone pole or stop sign makes a good test target. Suggested test range varies with manufacturer from about 200 feet (60 meters) to 500 feet (150 meters); some suggest the test target should be at the expected maximum operating range. The laser is swept pass the test target, a range reading should only occur when test target in reticle or aim circle area. To check vertical alignment the laser should be held at 90° (right or left 90 degrees). Also see chapter 6.4—Laser Radar Aiming / Alignment.
- check **range** on an object of known distance
  Most laser manufacturers recommend checking range accuracy by measuring the range of an object of known distance, some suggest measuring test targets at 2 different ranges. Recommended test range varies from about 250 feet (75 meters) to 500 ft (150 m); some suggest checking at expected minimum and maximum operating ranges. Radar

should range test target within specifications, typically plus or minus 0.5 ft (15 cm) to 2 ft (60 cm) depending on model.
- check ladar against **test target** (with a calibrated speedometer) of known speed (several speeds better)
- Insure ladar lab / shop calibration current (if state or local requirement)
  —ladar should be labeled with lab/shop calibration data (BY, DATE, DUE)
  —many agencies require ladars be lab/shop calibrated 1 or 2 times/year

Note, tuning forks do not register on ladar.

External optical surfaces (especially transmit/receive aperture, aim scope or aim display) should be periodically cleaned exactly as recommended by the manufacturer. Additionally all optical surfaces should be covered (capped) when not in use to prevent damage (scratches, pits, stains) resulting in performance degradation (detection range, beam alignment).

Beamwidth

Ladars have a major advantage over conventional microwave radars because lasers have much narrower beams, around 3 or 4 milliradians (0.17 or 0.23 degrees); a microwave beam is typically between 10 and 25 degrees. The sidelobes for laser systems are virtually nonexistent as compared to microwave. The figure below graphs the beamwidth spread for a 3, 3.5, and 4 milliradian (mR) beam. A 4 mR beam has good target selectivity to about 1500 feet (beam spreads about 6 feet) while the 3 mR beam has good target selectivity to about 2000 feet.

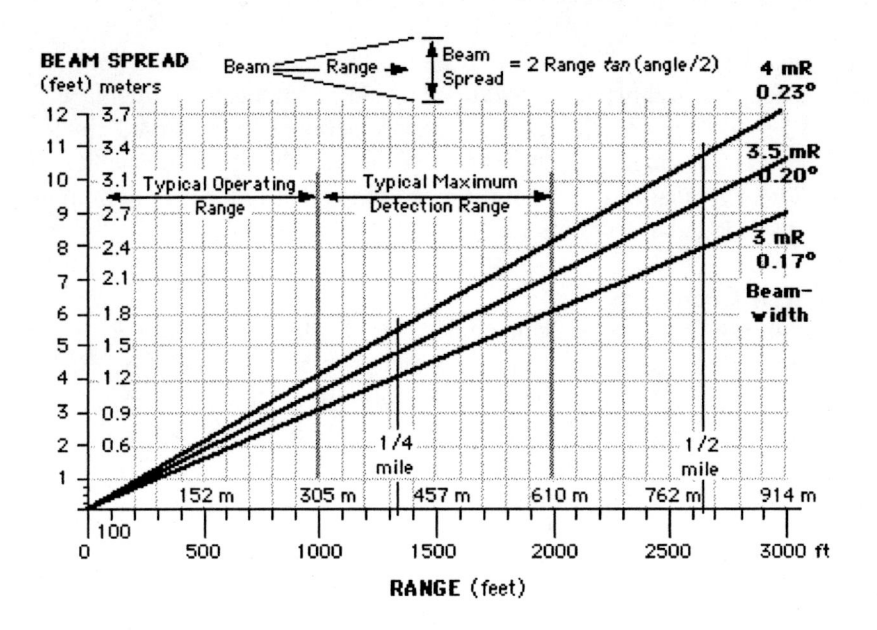

Traffic Radar Handbook
A Comprehensive Guide to Speed Measuring Systems

Figure 6.3-1—**Ladar Beamwidth(s)**

$$d = 2 R \tan (ß/2)$$
d = beam spread distance, R = range, ß = beamwidth

Operation
(not all operations available on all models)

Accuracy
Velocity: ± 1 mph (Accuracy decreases, error greater than 1 mph, above speeds of 60 to over 100 mph—depending on make/model). Range: ± 0.5 to ± 2 feet (varies with model).

Use
Ladars have narrow beams and must be aimed (a steady aim) with care; an operator cannot set up the ladar and wait for a violator, but must aim the ladar at a single target. Ladars use a scope (similar to a rifle scope) or a HUD (Heads Up Display) as an optical aiming device. A HUD allows the operator to simultaneously observe a target and see target data (speed and/or range) by

projecting a circle and target data onto a glass plate that the operator can see through.

Because ladar uses a narrow light beam, the ladar should not point through glass because the lightwaves may diffract and/or attenuate, reducing effective range. Flat clear glass will reduce ladar detection range; curved tinted glass reduces detection range greatly and may render the ladar unusable. Many windshields have coatings and/or tiny metal chips that reduce sunlight as well as severely degrade ladar detection range.

Atmospheric Conditions

Best performance when atmosphere cool, dry, and clear. IR energy propagation (ladar target detection range) is severely attenuated by H_2O in the form of fog, rain, snow, sleet, ice, and/or humidity. Dust particles, smoke, and carbon dioxide (CO_2) also reduce ladar target detection range. Note that atmospheric conditions (H_2O, dust particles, smoke, CO_2) do NOT affect ladar accuracy, only target detection range.

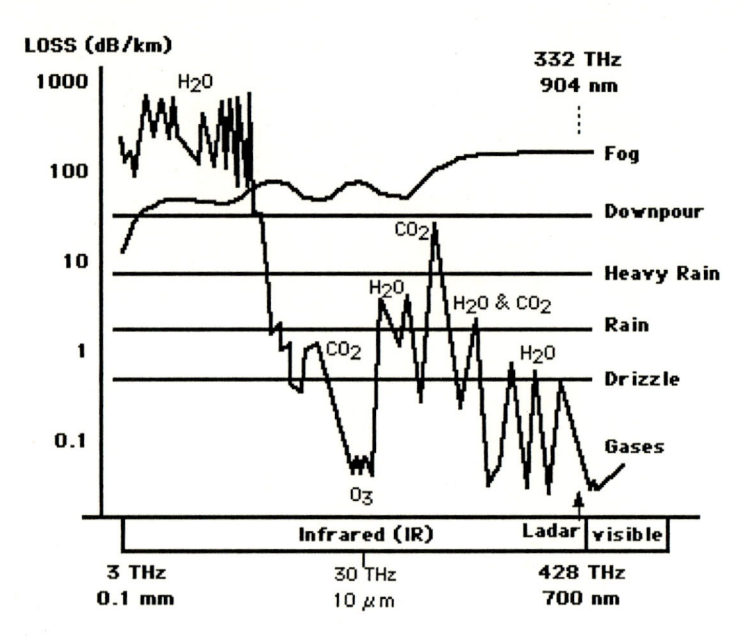

Figure 6.3-2—**Atmospheric Losses**

CONDITIONS: One-way attenuation (dB/km = dB/0.62 mi = dB/3281 ft = dB/1094 yds) in clear atmosphere (oxygen, and water vapor at 7.5 g/m^3), 1 atmosphere pressure and 293° K (68° F). Downpour—6 in/hr = 150 mm/hr,

Heavy Rain—1 in/hr = 25 mm/hr, Rain—0.2 in/hr = 5 mm/hr, Drizzle—0.01 in/hr = 0.25 mm/hr, Fog—0.1 g/m^3 (50 m = 164 ft visibility). Also see figure 8-3.2—Atmospheric Losses: Microwave to Light.

Most ladar manufacturers caution NOT to point the aperture directly at the sun. Direct sunlight has enough energy to damage most ladar receivers by burning out the light detector (diodes). Direct sunlight can also damage laser radar detectors. Also see chapter 6.5—Operational Problems with Laser Radar (Interference).

Configurations (power supplied by car battery and/or internal battery)
—Single unit—hand-held ladar gun
—Binoculars Style—hand-held

Instant On
Laser radars have narrow beams and only transmit when the operator has targeted a vehicle.

Velocity Measurement Modes
Under ideal conditions a ladar displays target speed in about 3/10 seconds or more (depending on model). Less than ideal conditions could force the ladar to take as long as several seconds or more, or miss a target entirely. In general the slower the target the more time required to calculate speed.

Stationary mode
Current ladars must operate from a stationary position only (no moving mode).

Target Direction
Most ladars have the option to measure approaching targets, receding targets, or both. Systems measuring traffic in both directions usually indicate the direction of the displayed target speed.

Range and Timing Modes
Ladars measure target speed but, because the speed measurement is based on target range (rate)—range information is also available. Ladars can also measure range of stationary targets (road signs, telephone poles, etc.). Knowing the range between two objects (or an object and patrol car) allows the operator to time targets to determine target speed. The range between two points is entered into the ladar, the operator pushes a button when the target passes between the two points and the ladar displays the calculated target speed (without transmitting).

Donald S. Sawicki

Detection Range

Ladars typically have a maximum detection range of about half a mile (about 2,600 feet or 800 meters) in good weather. Just like with microwave radar; target size, shape and reflectivity are major factors in ladar detection range performance. The front of a target vehicle will probably have a different (shorter) detection range than the rear. Some ladar manuals suggest an operator should aim the narrow ladar beam at the target license plate for best detection performance—the license plate is reflective and flat (most energy reflected back in direction of ladar). For reliable target detection, target ranges between about 50 and 600 feet (15 to 180 meters) work best.

Infrared ladar has many of the same proprieties as visible light. Red colors reflect ladar light better than violet or black colors; bright objects are better reflectors than dull objects. Chrome is a good reflector; white is a relatively good reflector for ladar. Flat horizontal objects reflect light back in the direction from which it came better than smooth objects that tend to scatter light. A vehicle can be made stealthy to ladar by removing the license plate and taping (black tape would probability work best) over all chrome parts and covering headlights, taillights, parking lights, and direction lights with tape (the less smooth the tape surface the better).

Chapter 6.4

Laser Radar
Aiming / Alignment

Ladars have narrow beams and must be aimed (a steady aim) with care; an operator cannot set up the ladar and wait for a violator, but must aim the ladar at a single target. A 3 or 4 milliradian (0.17 or 0.23 degree) beam must be aimed to within about 0.1 degrees. A narrow beam also makes a target difficult to track if the target is changing lanes, on a curve, or both. Also the greater the distance off the road the ladar (and the closer the target), the more movement required to track a target. While ladars are hand-held, it is easier to track a target with the ladar mounted to pivot from a fixed surface.

For the aim-spot to fall in the beam, the angle alignment error between the aim-spot and beam must be less than half of the beam width. The alignment error for a 3 mR beam must be less than 1.5 mR (0.086 degrees); for a 3.5 mR beam the error must be less than 1.75 mR (0.1 degrees); for a 4 mR beam the error must be less than 2 mR (0.11 degrees). Some ladars allow the operator to adjust sight alignment—both vertical (elevation) and horizontal (azimuth), some have factory adjustments only.

Most traffic ladars have the beam aperture offset slightly from the aiming device (typically about 3 inches, or about 7 centimeters). When the aim sight alignment is parallel to the beam center, the aim spot will be about 3 inches (aperture to aim offset) above the beam center for ALL ranges. If the aim sight is not parallel to beam center, the distance between aim spot and beam center varies with range (see figure below).

Donald S. Sawicki

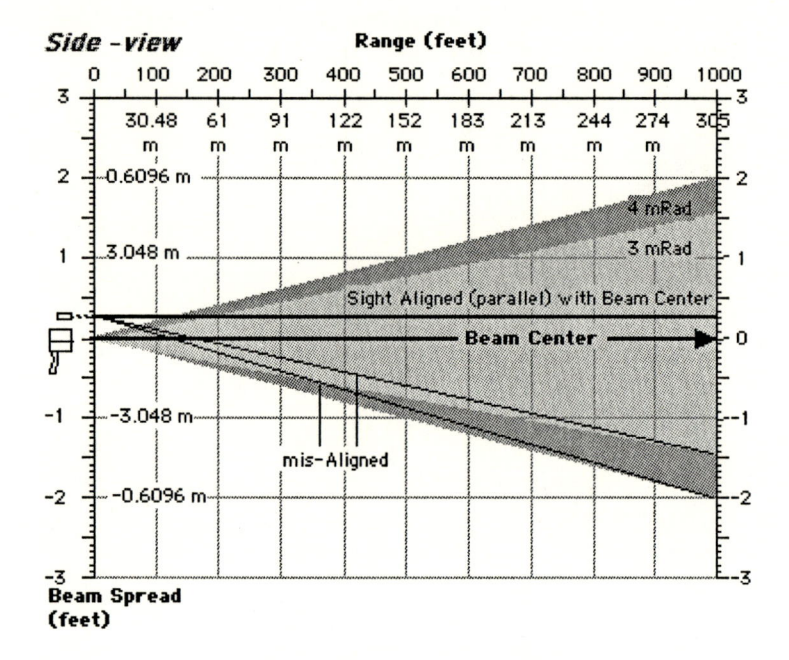

Figure 6.4-1—**Alignment**

Testing and/or beam alignment (horizontal and vertical) should be done at least once every day used. The greater the range between ladar and target, the more critical the aim and beam alignment. The range of the tested objects should be at the maximum range targets are expected. A ladar that appears aligned at 150 feet (45 meters) may be several feet off at 1,000 feet (305 meters), and on the beam edge (see figure above).

The greater the ladar angle error (between aim and beam center) and the greater the target range, the larger the distance between aim spot and beam center. An angle error of 0.25 degrees translates to about 6 feet (2 meters) difference between aim spot and beam center at 1/4 mile (0.4 km) range; and about 11 feet (3 meters) difference at a range of 2,450 feet (745 meters).

Distance Between expected Aim Spot and Beam Center
= **Range x *tan* (angle error)**

The ladar must be aimed to within tenths of a degree, the aperture must be aligned within hundredths of a degree to the optical aiming device. A two-aperture (separate transmit and receive) ladar must align both apertures to thousandths of a degree, and this combination must then be aligned to the optical aiming device. Even if a ladar is perfectly aligned an angle (aim) error may be introduced if the ladar operates behind a windshield (windshield may diffract signal).

Donald S. Sawicki

Chapter 6.5

Operational Problems with Laser Radar

Outline
- Cosine Effect
- Target Speed / Acceleration
- Interference
- Scanning Error

The Cosine Effect
The Cosine Effect degrades ladar speed measurement accuracy exactly the same way it degrades microwave radar accuracy.
see chapter 4.1—Cosine Effect Geometry
see chapter 4.2—Cosine Effect on Measured Speed

Target Speed / Acceleration
Laser radars typically sample target range in time intervals of 0.3 to 1 second or more. Slow targets may require more time to establish speed. For example a 10 mph target travels at 14.7 feet/second, or about 4.4 feet in 0.3 seconds (minimum ladar track time); a 5 mph target travels at 7.3 feet/second, or about 2.2 feet in 0.3 seconds. If target reflections vary (between front bumper and passenger compartment) more than target range changes, speed measured may be grossly inaccurate or difficult to establish.

Targets accelerating or decelerating (due to the cosine effect or not) more than about 1 mph (or 1 kmh) per ladar cycle (integration or sample time) may be difficult or impossible for a ladar to track. The ability to track a target depends on the target acceleration or deceleration component, ladar integration time (typically 0.3 to 1 second or more), and ladar speed tolerance (typically 1 mph or 1 kmh per cycle).

LADAR ACCURACY mph or kmh or knots...	LADAR SAMPLE TIME	MAXIMUM SPEED CHANGE mph or kmh or knots... PER SECOND	g's mph	kmh	knots
+/- 1	2000 ms	0.5	0.02	0.01	0.03
"	1000 ms	1	0.05	0.03	0.05
"	800 ms	1.25	0.06	0.04	0.07
"	500 ms	2	0.09	0.06	0.10
"	360 ms	2.8	0.13	0.08	0.15
"	333 ms	3	0.14	0.09	0.16
"	300 ms	3.3	0.15	0.09	0.17

Table 6.5-1—**Maximum Acceleration Given Sample Time**

ms is milliseconds (1000 ms = 1 sec)

Also see chapter 5.4—Target Acceleration (Deceleration) for more details.

Interference

If for any reason intended signal returns are **interrupted**, the ladar may not be able to determine target velocity (for that sample time). If another vehicle passes between the intended target and ladar, or if the ladar were to try to scan through a tree, branches, sign, utility pole or tower, some or all target returns could be missed. See appendix F—Ladar Technical Details for more information on signal returns.

Bright lights (such as halogens) beaming directly into a ladar aperture (beam) may desensitize or entirely mask target echoes. The degree of interference depends on the intensity and wavelengths of the light. The brighter and closer the light source, the greater the chance of interference.

Light (especially white light) consists of multiple wavelengths (frequencies) with random phases spread across a relatively wide part of the spectrum (not just the visible part necessarily). The lights' spectrum depends on and varies with source. If the light detector in the ladar receiver responds to some or all of an interfering light's spectrum and the light is intense enough, ladar target detection range may be reduced. Reflections from objects illuminated by the sun (large bandwidth and very intense) could also lessen a ladar's performance. Also see chapter 6.3—Laser Radar Operation (Atmospheric Conditions).

Donald S. Sawicki

Scanning Error

Because the ladar measures range difference to compute speed, if the ladar is scanned in a manner that causes signal returns (such as the ground or any roughly continuous surface like a guardrail) to decrease (or increase) in range, a false speed reading is possible. The magnitude of the false reading is directly proportional to the change in range for one sample period (see figure below). For example a ladar with a **0.333 second** (333 ms) sample time may register a speed of **100 mph** (161 km/Hr) when the ladar scans a guardrail a distance of approximately **50 feet** (say 500 to 450 feet) in 0.333 seconds.

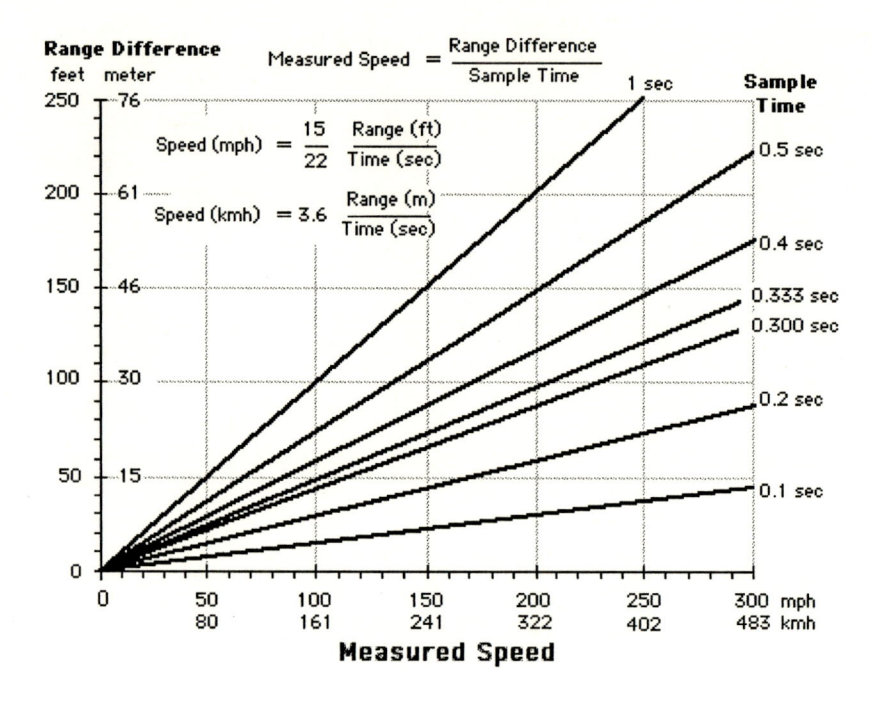

Figure 6.5-1—**Range vs Speed / Sample Time**

A scanning error is possible if beam spread is wide enough (or beam to cross-hairs alignment error great enough) to include the ground, typically at longer ranges. If ground echoes are stronger than target echoes—any radar ANGULAR movement that shifts ground echo range could translate into a measured speed of ground echo range difference (during radar sample time).

Traffic Radar Handbook
A Comprehensive Guide to Speed Measuring Systems

Chapter 6.6

Laser Radar Speed Error (due to range error)

Outline
- Target Reflection varies between Rear—Front of Target
- Range Resolution
- Range Variations

Because ladar uses range (change in range over time) to compute velocity, range errors directly affect velocity computations. Ladars typically transmit tens to hundreds of pulses per sample period which means tens to hundreds of target range measurements used to determine target speed (one speed measurement).

Ladar pulse width determines range resolution, which is the ladar's ability to separate closely spaced targets in the range dimension (as opposed to the angular dimension). To resolve two targets in range, the targets must be separated by at least half the transmit pulse width. Note that light travels about 1 foot per nanosecond (1 ft / ns). A ladar with a pulse width of 35 ns cannot distinguish targets that are separated by less than 17.5 ns (about 17.5 feet or 5.3 meters). A 5 ns pulse (about 2.5 feet or 0.76 meters range resolution) could result in some echo returns from the front of a target and some returns from the rear of the target (depending on target shape and geometry).

Target Reflection varies between front—rear of Target
If a laser radar tracks a target such that the beam moves from the back of the target toward the front of the target (in a continuous manner) during 1 sample period—measured speed is high. When the change in target range measures long, the velocity calculates high.

If a laser radar tracks a target such that the beam moves from the front of the target toward the back of the target (in a continuous manner) during 1 sample period—the measured speed is low. When the change in target range measures short, the velocity calculates low.

Donald S. Sawicki

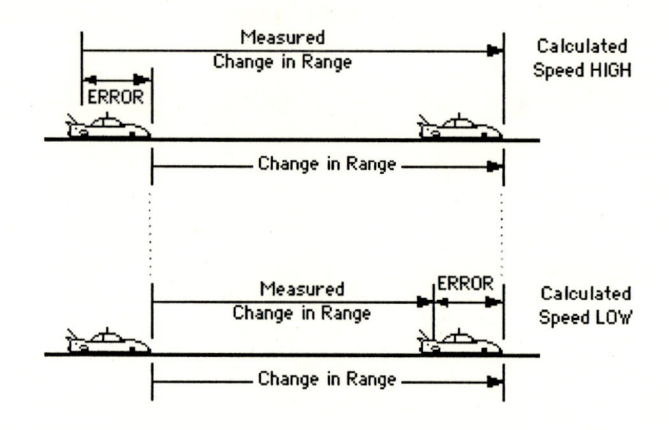

Figure 6.6-1—**Change in Range Error**

When the range error adds to target change in range (during ONE sample time), velocity measures high; when the range error subtracts from target change in range, velocity measures low. For either an approaching or receding target a beam moving from the back to the front of the target adds to actual target speed; a beam moving from the front to the back subtracts from target speed.

The speed error is a function of ladar sample time and range error as shown in the table and figure below. The larger the range error (during a sample time), the larger the velocity error. A minor range error of a foot translates to a 2 mph error for a ladar with a sample time of 1/3 seconds (0.333 sec).

Sample Period (sec)	0.1 sec	0.2 sec	0.3 sec	0.333 sec	0.4 sec	0.5 sec
RANGE ERROR			VELOCITY ERROR			
0.5 ft	3.41 mph	1.70 mph	1.14 mph	1.02 mph	0.85 mph	0.68 mph
1.0 ft	6.82 mph	3.41 mph	2.27 mph	2.05 mph	1.70 mph	1.36 mph
1.25 ft	8.52 mph	4.26 mph	2.84 mph	2.56 mph	2.13 mph	1.70 mph
1.5 ft	10.23 mph	5.11 mph	3.41 mph	3.07 mph	2.56 mph	2.05 mph
metric						
0.5 meters	18 kmh	9 kmh	6 kmh	5.4 kmh	4.5 kmh	3.6 kmh
1.0 meters	36 kmh	18 kmh	12 kmh	10.8 kmh	9 kmh	7.2 kmh

Table 6.6-1—**Speed Error for given Unit Range**

Traffic Radar Handbook
A Comprehensive Guide to Speed Measuring Systems

A range error of plus 0.5 foot during a 0.333 second sample period translates to a measured speed 1.02 mph higher (or lower) than actual. The velocity error is directly proportional to range error (see figure below). A range error of 2 feet (4 x 0.5) during a 0.333 second sample period translates to a speed error of 4.08 mph (4 x 1.02).

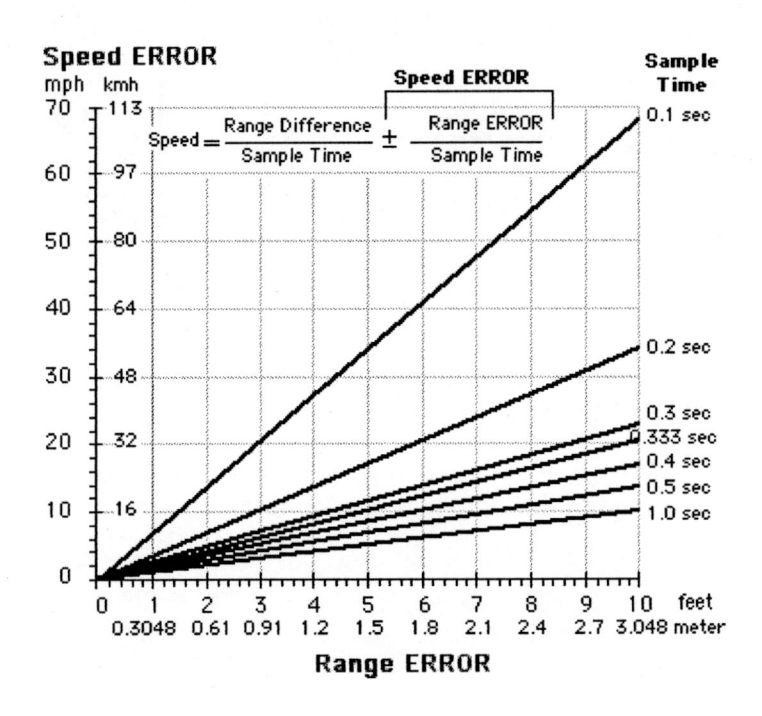

Figure 6.6-2—**Range Error vs Measured Speed Error**

Range Resolution

Ladar pulse width determines range resolution, which is the ladar's ability to separate closely spaced targets in the range dimension (as opposed to the angular dimension). To resolve two targets in range, the targets must be separated by at least HALF THE TRANSMIT PULSE WIDTH.

Donald S. Sawicki

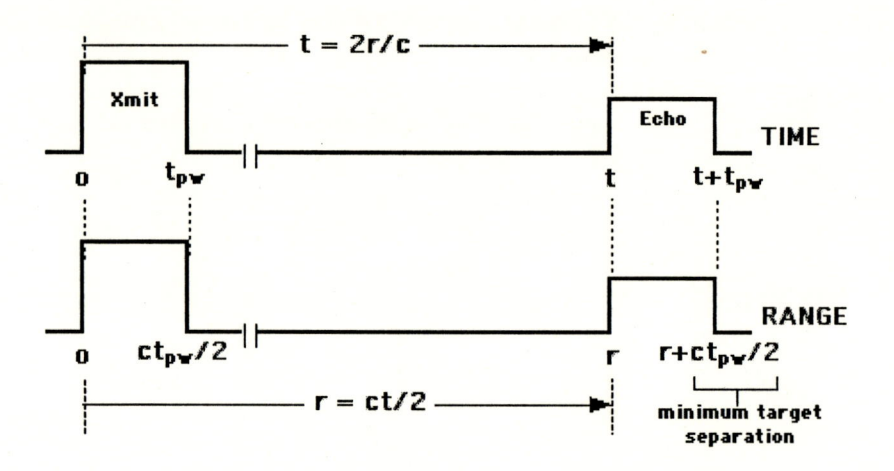

Figure 6.6-3—**Minimum Target Separation**

Time is round trip pulse travel time (from ladar to target and back to ladar). Range is actual distance between ladar and target. Note in the RANGE dimension the trailing edge of the echo pulse represents the minimum range to distinguish the next echo. Minimum range separation for ladar to distinguish 2 targets is;

$$R > c\, t_{pw} / 2$$

R = range separation
c = speed of light
t_{pw} = pulse width (time)

Range Separation in feet > **0.4918 x Pulse Width** in ns
Range Separation in meters > **0.1499 x Pulse** Width in ns

ns = nanoseconds = 0.000 000 001 seconds

Traffic Radar Handbook
A Comprehensive Guide to Speed Measuring Systems

PULSE WIDTH	MINIMUM RANGE SEPARATION		PULSE WIDTH	MINIMUM RANGE SEPARATION	
1 ns	0.49 ft	0.15 m	51 ns	25.08 ft	7.64 m
2 ns	0.98 ft	0.30 m	52 ns	25.57 ft	7.79 m
3 ns	1.48 ft	0.45 m	53 ns	26.06 ft	7.94 m
4 ns	1.97 ft	0.60 m	54 ns	26.56 ft	8.09 m
5 ns	2.46 ft	0.75 m	55 ns	27.05 ft	8.24 m
6 ns	2.95 ft	0.90 m	56 ns	27.54 ft	8.39 m
7 ns	3.44 ft	1.05 m	57 ns	28.03 ft	8.54 m
8 ns	3.93 ft	1.20 m	58 ns	28.52 ft	8.69 m
9 ns	4.43 ft	1.35 m	59 ns	29.02 ft	8.84 m
10 ns	4.92 ft	1.50 m	60 ns	29.51 ft	8.99 m
11 ns	5.41 ft	1.65 m	61 ns	30.00 ft	9.14 m
12 ns	5.90 ft	1.80 m	62 ns	30.49 ft	9.29 m
13 ns	6.39 ft	1.95 m	63 ns	30.98 ft	9.44 m
14 ns	6.88 ft	2.10 m	64 ns	31.47 ft	9.59 m
15 ns	7.38 ft	2.25 m	65 ns	31.97 ft	9.74 m
16 ns	7.87 ft	2.40 m	66 ns	32.46 ft	9.89 m
17 ns	8.36 ft	2.55 m	67 ns	32.95 ft	10.04 m
18 ns	8.85 ft	2.70 m	68 ns	33.44 ft	10.19 m
19 ns	9.34 ft	2.85 m	69 ns	33.93 ft	10.34 m
20 ns	9.84 ft	3.00 m	70 ns	34.42 ft	10.49 m
21 ns	10.33 ft	3.15 m	71 ns	34.92 ft	10.64 m
22 ns	10.82 ft	3.30 m	72 ns	35.41 ft	10.79 m
23 ns	11.31 ft	3.45 m	73 ns	35.90 ft	10.94 m
24 ns	11.80 ft	3.60 m	74 ns	36.39 ft	11.09 m
25 ns	12.29 ft	3.75 m	75 ns	36.88 ft	11.24 m
26 ns	12.79 ft	3.90 m	76 ns	37.38 ft	11.39 m
27 ns	13.28 ft	4.05 m	77 ns	37.88 ft	11.54 m
28 ns	13.77 ft	4.20 m	78 ns	38.36 ft	11.69 m
29 ns	14.26 ft	4.35 m	79 ns	38.85 ft	11.84 m
30 ns	14.75 ft	4.50 m	80 ns	39.34 ft	11.99 m
31 ns	15.25 ft	4.65 m	81 ns	39.83 ft	12.14 m
32 ns	15.74 ft	4.80 m	82 ns	40.33 ft	12.29 m
33 ns	16.23 ft	4.95 m	83 ns	40.82 ft	12.44 m
34 ns	16.72 ft	5.10 m	84 ns	41.31 ft	12.59 m
35 ns	17.21 ft	5.25 m	85 ns	41.80 ft	12.74 m
36 ns	17.70 ft	5.40 m	86 ns	42.29 ft	12.89 m

Donald S. Sawicki

37 ns	18.20 ft	5.55 m	87 ns	42.79 ft	13.04 m
38 ns	18.69 ft	5.70 m	88 ns	43.28 ft	13.19 m
39 ns	19.18 ft	5.85 m	89 ns	43.77 ft	13.34 m
40 ns	19.67 ft	6.00 m	90 ns	44.26 ft	13.49 m
41 ns	20.16 ft	6.15 m	91 ns	44.75 ft	13.64 m
42 ns	20.65 ft	6.30 m	92 ns	45.24 ft	13.79 m
43 ns	21.15 ft	6.45 m	93 ns	45.74 ft	13.94 m
44 ns	21.64 ft	6.60 m	94 ns	46.23 ft	14.09 m
45 ns	22.13 ft	6.75 m	95 ns	46.72 ft	14.24 m
46 ns	22.62 ft	6.90 m	96 ns	47.21 ft	14.39 m
47 ns	23.11 ft	7.05 m	97 ns	47.70 ft	14.54 m
48 ns	23.61 ft	7.20 m	98 ns	48.19 ft	14.69 m
49 ns	24.10 ft	7.34 m	99 ns	48.69 ft	14.84 m
50 ns	24.59 ft	7.49 m	100 ns	49.18 ft	14.99 m

Table 6.6-2—**Pulse Width versus Range Separation**

Note that light travels about 1 foot per nanosecond (1 ft / ns); ladar pulse round trip travel time for a target 1 foot away takes 2 ns.

A ladar with a pulse width of 35 ns cannot distinguish targets separated by less than about 17 feet (5.3 meters).

A 5 ns pulse (about 2.5 feet or 0.76 meters range resolution) could result in some echo returns from the front of a target and some returns from the middle or rear of the target (introducing a range error).

Range Variations

If during a sample time some pulse echoes reflect from the front of a target and others from the middle or rear, the ladar may not be able to compute target speed if the range measurements (during 1 sample period) do not correlate. If the beam spot (on the target) is small compared to target size (typically at ranges less than 500 ft or 150 m) any operator (aiming) movement could introduce range errors as described above. At longer ranges where the beam illuminates most or all of the target, range resolution becomes a factor.

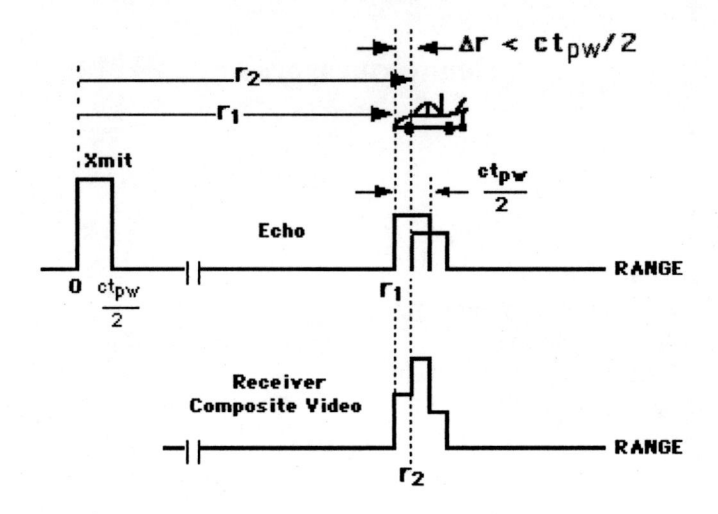

Figure 6.6-4—**Target Echo Delay**

Donald S. Sawicki

Chapter 7.1

Countermeasures

Outline
- Scams
- Driving Tips
- Radar Detectors
  —Safety Warning System option
  —Safety Alert option
- Radar Detector Detectors (RDD)
  —VG-2 Interceptor
  —Stalcar
- Microwave Jammers
- Laser Jammers

Also see chapter 6.3—Laser Radar Operation (Detection Range).

Scams

One of the first attempts to fool traffic radar was to hang reflective objects on a vehicle; these included dangling loose chains, putting foil or steel marbles in hubcaps and securing strips of foil to the radio antenna. None of these counters did anything to fool traffic radar and, in fact, may have enhanced the target's radar cross-section (echo), enabling the radar to detect the target at longer range! Even so, drivers that used these techniques were subject to arrest in Rochester, New York, in 1953.

Today products exist that CLAIM to interfere with microwave traffic radar. Be especially skeptical of jammer claims, and in particular any **passive jammers** (all known passive jammers are useless). Some test indicate not only do these jammers not work, but they may enhance the target's radar cross-section (a jammer antenna, alone, tuned to microwave frequencies is a great reflector) enabling the radar to detect the target at longer range (10 to 30 percent in some cases!). Even so, drivers that use jammers are subject to FCC penalties (FCC public notice 1996). Also see Microwave Jammers below.

Some **hood covers** (bras) and **license plate covers** are suppose to absorb and/or disperse microwave radar energy thus reducing target radar cross-section and radar detection range. Any reduction is small with only a slight (**if any**) degradation in radar performance (maximum detection range).

Traffic Radar Handbook
A Comprehensive Guide to Speed Measuring Systems

Some **license plate covers** supposedly disperse laser light denying a laser radar target reflections from the license plate. However if the laser is not aimed at the (covered) license plate or the beam is wide enough to overlap other parts of the vehicle this method will fail. In fact most reports indicate license plate covers (alone) have little if any effect on laser (or microwave) radar performance.

False information has circulated on the Internet (mainly through Email in 2000 and 2001) concerning avoiding **points** against your driver's license due to a traffic ticket (speeding, red light running, etc.) issued in any U.S. state. Basically the scam message falsely claims by over paying a ticket by a few dollars (mail-in only), then not cashing a check for the difference when the system corrects the overpayment, prevents the system from assessing points against your driver's license (until the difference check is cashed)—THIS IS NOT TRUE and DOES NOT WORK!

Driving Tips

If one knows of or is warned (from a **CB** or a **passing motorist flashing headlights**) of an impending speed trap, it's a good idea to drop speed five mph below the posted limit. This may or may not prevent a ticket, but could help in court if a ticket is issued. If one can claim the vehicle speed was at least five mph below the limit, target speedometer accuracy will not be as questionable. It's not a bad idea to have the speedometer calibrated by a qualified shop, even after a ticket, so one may determine if the speedometer was accurate at the time of the alleged violation. If one has the speedometer calibrated be sure to retain all documentation for court evidence.

If a driver believes (such as a warning from a radar detector or a visual sighting) a traffic radar has just started illuminating and/or tracking the driver's vehicle; a good countermeasure is to **gently decelerate** (not a dead skid) about *3 or 4 mph per second (3 or 4 kmh per second)*. A traffic radar/ladar may experience problems measuring targets decelerating (or accelerating) because speed is changing too fast.

Some police switch on the radar only when the operator is ready to target a vehicle. Some radars transmit only long enough (tens of to a second or more) to get a speed reading. With Across the Road radar (photo radar) a target is in the main beam for a short time (tens of to 2 seconds). Depending on the situation a driver may only have a fraction of a second to respond, and in some cases may not be warned at all until after the radar gets a measurement. **The lesson: do not depend solely on a radar detector to warn a traffic radar is in the area.**

Laser radars, ladars, have a narrow beam making it difficult to detect unless the laser is aimed directly at a laser detector (not just the vehicle). Laser radars could (under good conditions) measure target speed in about 1/3 of a second giving the target vehicle little or no time to react. A laser detector may sense a laser signal reflecting off another target or a stationary object, but this condition usually has a short duration if it occurs at all. Laser radar detectors are generally not as effected as microwave detectors.

If pulled over most officers prefer that you stay in your vehicle. When the officer approaches be polite and keep your hands visible—at night turn on your interior doom light before the officer approaches. If **ticketed** collect as much information as possible about the circumstances. Besides information on the ticket some or all of the ADDITIONAL DATA below may be helpful when analyzing events later.

ADDITIONAL INFO (if ticketed)
---

(1) RADAR or LADAR
   Make / Model: _____ / _____
   Sample Time: _____ milliseconds (ms)
   Beamwidth: _____ (AZ) x _____ (EL) degrees

(2) RADAR Off Road (target lane): _____ ft m

(3) TRACKING RANGE and/or TIME (to even)
   Start: _____ ft m _____ sec
   End: _____ ft m _____ sec
   TOTAL: _____ ft m _____ sec

(4) OTHER TARGETS with respect to intended target
   front   -   side(s)   -   rear

(5) OTHER TRANSMITTERS close to radar
   (note all patrol car antennas)

---

(6) MOVING RADAR Patrol Speed: _____ mph kmh
   Radar Angle to patrol car: _____ degrees

Traffic Radar Handbook
A Comprehensive Guide to Speed Measuring Systems

Radar Detectors
ECM—Electronic Countermeasures

Some radar detectors will identify a radar signal sooner than others, but the difference is usually not significant. If a traffic radar is continuously transmitting, a good detector can give a mile or two warning under ideal conditions. Bends in the road or objects that mask the radar signal will reduce radar detector performance as well as radar performance.

One can get the most from any given radar detector by proper installation—the detector antenna should be clear of any obstructions (especially metal such as wiper arms) and mounted as high as possible. Some **electrically heated windshields**, such as the Ford Instaclear or General Motors (GM) Electriclear, have metal film coatings reflective to microwave signals. Not only do these windshields obstruct signals and prevent the use of a radar detector (or a radar), these windshields increase a vehicle's radar cross section (RCS) increasing radar detection range.

A large selection of traffic radar detectors is available with a variety of options. Some typical options are listed below, not all detectors have all functions.

Signals Detected
- **X** band radars (10.525 GHz ± 25 MHz)
- **K** band radars (24.025 - 24.50 GHz)
- **Ka** band radars (33.4 - 36 GHz)
- **wideband Ka** radars (frequency hops 33.4 - 36 GHz)
- **laser**—laser radar, ladar or lidars (904 nm Infrared)

- **Safety Warning System** (SWS) messages (see below)
- **Safety Alert** Signals (see below)

- **VG-2** radar detector detector (see below)
- **Stalcar** radar detector detector (see below)

- **Strobe Alert**—detects strobe light used by emergency vehicles to change stoplight green.

Countermeasures
- Immunity from VG-2 radar detector detector (RDD)
- Immunity from Stalcar radar detector detector (RDD)
    Immunity—different LO's (see below), or detects RDD LO and shuts down.

- Built-in or hidden antenna(s) / processor / display

Antenna(s)
- front
- Front and Rear
- 360 degree coverage (reflections off other targets)

Indicators
- **Signal strength**
  light(s) and/or LCD—the stronger the signal the more lights (or bars) and/or rapid the blinking audio beeping—the stronger the signal the more rapid the beeps
- **Signal type**—X, K, Ka, Wideband Ka, or laser radar; RDD; Safety signal. Some models measure and display frequency for X, K, and Ka radar transmissions.
- Safety Message display
- Low Voltage
- synthesized voice—signal type and/or safety message etc.

Controls
- On / Off
- self-test
- audio volume
- mute
- display brightness adjust
- city / highway (city setting reduces detection range)

Power Source
- vehicle power source—via cigarette lighter or DC plug to connector / fuze / cable
- internal battery

Navigation and Additional Features
- compass
- altimeter—altitude above sea level
- GPS—latitude / longitude using Global Positioning System
- weather radio—receives NOAA weather radio (VHF) broadcast
- temperature—measures outside temperature
- RU Alert—requires driver to interact with detector to keep driver alert (awake)

If a driver needs an RU Alert, the driver should get off the road and rest.

Safety Warning System (SWS)

In addition to detecting traffic radar signals, Safety Warning System (SWS) detectors (available since 1996, approved by the FCC on 1999 Jan 28, FCC Rules Part 90.103) are capable of receiving specific messages broadcast from special transmitters (fixed or mobile). Besides looking for traffic radar signals, SWS capable receivers also look for safety warning signals (once every 3 to 6 seconds). A special transmitter broadcast a coded signal (a safety message) intended for reception by motorist in the area. An SWS capable detector alerts the driver with an audio tone that a message has been received and displays the message (some detectors use a synthesized voice). Sixty-four short fixed text messages are possible, up to 2 different messages may be broadcast one after the other. Variable messages are allowed if none of the available fixed text messages is appropriate. Radar detectors that do not have the SWS feature will detect a SWS signal message as a K band radar.

Transmitters broadcast at 24.1 GHz in the (lower part of) traffic radar K band. Transmitters function in both mobile (emergency vehicles such as police fire or ambulance, slow and/or oversized vehicles, school buses) and stationary modes (construction areas, dangerous stretches of road, or area information). For some stationary transmitters the broadcast message can be programmed from a remote location via radio or phone link.

SWS Transmitter
Frequency 24.1 GHz ± 50 MHz
Power (ERP) 50 milliwatts (mW)
Beamwidth 23 degrees (horizontal)

Georgia Tech Research Institute (GTRI) developed the system specifications as part of the Smart-Highway plan. Safety Warning System, L.C. is an organization formed to develop and promote the system, and has donated transmitters to local and state governments and at least 3 countries (Russia, New Zealand, and the Netherlands) as of 1998.

SWS LICENSEES
BEL-Tronics, L.L.C.
Escort, Inc.
M.P.H. Industries, Inc., first company to manufacture SWS transmitters
SK Global America, Inc.
Santeca Electronics, Inc. (formerly Sanyo Technical)

Donald S. Sawicki

Star Dreams Corp., formed 1992, Rahway, NJ, starting in 1998 markets transmitters to Russia and former Block nations
Uniden America Corp.
Whistler Corp.
Yupiteru Industries Co., Ltd

Safety Alert option
Cobra Electronics Corp.

Safety Alert System, introduced in the 1990's, transmits 1 of 3 fixed messages (**Emergency Vehicle—Train—Road Hazard**) intended for reception by special receivers or radar detectors (designed to receive the Safety Alert signal). Radar detectors that do not have the Safety Alert feature will detect the signal message as a K band radar. The Safety Alert System transmits at 24.07 GHz, 24.11 GHz, and 24.19 GHz.

Radar Detector Detectors (RDD)
ECCM—Electronic Counter-Countermeasures

In some states in the U.S., some provinces in Canada, and some states in Australia radar detectors are illegal to use and/or possess, with the owner subject to fines and/or confiscation of the device. In the United States Federal Highway Administration regulations, as of January 1994, prohibit radar/laser detector use in vehicles over 10,000 pounds.

A radar detector detector (RDD) is used to enforce anti-detector laws by sensing signals that leak from a radar detector's local oscillator (LO); detecting a signal indicates a radar detector is operating (on) and near-by. All superheterodyne receivers have a local oscillator, and most if not all radar detectors use a superheterodyne receiver. Many radar detectors use a local oscillator that operates around **11.558 GHz**. Some radar detectors have an LO that operates around **12 to 15 GHz** (intended to defeat some radar detector detectors).

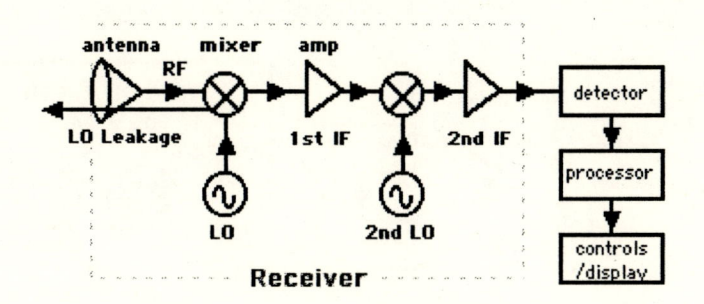

Figure 7.1-1—**Radar Detector Local Oscillator (LO) Leakage**

A superheterodyne receiver coverts RF (radio frequency) signals to an intermediate frequency (IF) by mixing the RF signal with the LO signal (IF = RF ± LO). Mixers are not perfect and leak some of the LO signal to (and out) the antenna. Some radar detectors may have a pre-amp or isolator between the mixer and antenna that should reduce LO leakage, but may not eliminate it. The old (and unreliable) crystal radar detectors (X band) cannot be detected by radar detector detectors (RDDs) because the crystal radar detectors do not have a local oscillator.

VG-2 Interceptor
Radar Detector Detector (RDD)
Technisonic Industries Ltd.
Missasauga, Ontario (Canada)

The VG-2 Interceptor radar detector detector (introduced in the 1980s) measures microwave signals in the band 11.4 - 11.6 GHz and is suppose to have detection ranges from 1/4 mile to 2 miles depending on the strength of radar detector LO leakage. Because the VG-2 has only one IF (one frequency conversion), the device is more susceptible to false alarms than a double frequency conversion receiver. Some radar detectors are capable of detecting the VG-2 LO leakage (a radar detector detects the radar detector detector).

Donald S. Sawicki

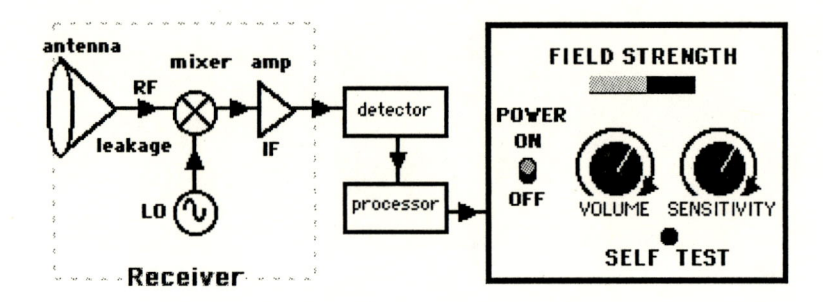

Figure 7.1-2—VG-2 Radar Detector Detector

Some **mobile amateur radio** transceivers leak signals in the X band that set off the VG-2. Several people operating legal ham radios were wrongfully accused of possessing radar detectors because their amateur radios were detected by a radar detector detector. **Microwave point-to-point** communication systems operate in the band **10.7 to 13.25 GHz.** These systems have the potential to register on a radar detector detector if the microwave transmit frequency is in or near the radar detector detector receiver band (11.4 to 11.6 GHz for the VG-2).

Stalcar
Radar Detector Detector (RDD)
Stealth Microsystems Pty Ltd.
Brisbane, Queensland (Australia)

Stalcar (introduced Feb 2000) is a radar detector detector (RDD) intended to pick up a variety of radar detectors. An audio alarm sounds when a signal is detected, and a bar graph display indicates signal proximity (range based on signal strength). The Stalcar is multi-band and uses a double down conversion receiver. The unit mounts to (inside) windshield or side window using suction cups. Detection range reported to be as high as 1 kilometer (0.6 miles) in some instances.

INDICATORS:	audio alarm (signal detected)
	graph display of signal proximity (range)
RECEIVER:	double down conversion superheterodyne
	varactor tuned Gunn Oscillator
EXTERNAL POWER:	13.50 to 14.50 Vdc, 0.5 amps max
TEMPERATURE RANGE:	-5 to 65 °C (23 to 149 °F)
WEIGHT:	475 grams (17 oz—just over 1 lb)
DIMENSIONS (LxWxH):	155 x 93 x 35 mm (6.1 x 3.7 x 1.4 in)

Microwave Jammers
ECM—Electronic Countermeasures

From time to time traffic radar jammers appear in magazine ads and such (the Internet). Be aware MOST, if not ALL, of these jammers are useless (absolutely no effect at any range under any conditions) based on test reports by Automobile magazine, Car and Driver, RADAR Reporter and Truckers News. The legality of jammers is also in question and somewhat up to the whims of bureaucrats and politicians (local, state, and federal) regarding enforcement and court interpretation of the law. Several U.S. states and some countries prohibit the use or possession of a jammer.

For the United States **FCC** (Federal Communications Commission) to consider an intentional radiator legal the field intensity (power) must meet FCC limits (Rules Part 15) **AND** the device must perform some function for the public good. Traffic radar jammers are not considered good for the public by the FCC.

The FCC considers the use of traffic radar jammers as malicious interference and strictly prohibited by the Communications Act of 1934, as amended, as well as by FCC rules. Anyone using a jammer risks such penalties as losing FCC licenses, paying a fine, or facing criminal prosecution (from FCC Public Notice, DA 96-2040, 1996 DEC 9).

Passive jammers are suppose to re-radiate the radar signal after distorting it (adding noise and/or rapidly shifting frequency) in such a way the true target reflection is masked by the distorted signal. A passive jammer does not generate or amplify a signal, only channel or redirect the radar signal (after distorting) back toward the radar. For this method to work the jammer (distorted signal) power must be as large as or greater than target reflected power—the jammer antenna would need to capture well over half of all the radar energy striking the

Donald S. Sawicki

target (a very large jammer antenna), and be aligned to the radar antenna. To date all known passive jammers have absolutely no effect on any radar under any circumstances.

On 4 December 1997 the FCC ruled passive jammers violate federal regulations because the jammers radiate RF (Radio Frequency) energy that (or is intended to) adversely affect the ability of law enforcement officials to protect public safety on the highways. The ruling was based on a passive jammer (Sprint II) made by Rocky Mountain Radar (out-of-business) in Colorado. Before this ruling passive jammers were not considered transmitters and thus not covered by FCC regulations.

Active jammers either detect a radar signal before transmitting (jammer must have fast reaction time), or continuously transmit whether a radar signal is present or not (less sophisticated jammers). Many traffic radars can detect jamming signals (alerting the radar operator) even when the radar is not transmitting, this is why a jammer should only transmit when a radar signal is present.

A variety of jamming signals, depending on jammer, are used to blind (noise jamming) or fool (deception jamming) traffic radar. Transmitting brute force noise is one technique used to obscure target echoes; however this may be detected by many radars as a jammer signal alerting the operator. Another technique is to transmit a signal (fake target echo) at a frequency that will get into the radar Doppler discriminators/filters. The signal appears to the radar as a legitimate target overriding the real target echo; then the jammer signal will drift in frequency slow enough for the radar to track and too fast/erratic for the radar to calculate speed (effectively blinding the radar with no indication to the operator). The jammer signal could also force the radar to a set predetermined frequency that causes the radar to read a speed (set by the jammer). These last two techniques depend on knowing the EXACT frequency transmitted by the radar and generating exact (very tight tolerance) frequencies for transmission.

Most traffic radars have some ECCM (Electronic Counter-Countermeasures) capabilities for detecting, alerting the operator, and countering Radio Frequency Interference (RFI) and jamming signals. The degree of success a radar has detecting and countering an unwanted signal varies with unwanted signal parameters, range between transmitter and radar, and radar make and model.

To date only one known (active) jammer actually works; the Stealth/VRCD made by Stealth Technologies in Naperville, Illinois—reported (June 1998) to be out of business due to an FCC crackdown on radar jammer manufacturers. This jammer counters X and K band traffic radars; some later models reportedly also counter Ka band (fixed frequency) radars. The unit has an audio alert and LED to indicate when a radar is detected, and another LED to indicate the jammer is transmitting. The unit reportedly uses a fake target echo to blind the radar (as described in the above paragraph). This jammer received high marks from Car and Driver magazine, RADAR Reporter, and Truckers News. The RADAR Reporter (1993 NOV edition) tested the Stealth/VRCD against 5 different radars; 4 out of 5 radars could not register target speed at all. In all cases the radars gave no indication a jamming signal was present. The Decatur MV-715 (X band) managed to measure target speed (burn-through the jamming) on one test run, at a mere 150 foot range (will vary with target radar cross section). The jammer managed to detect and defeat the traffic radar in plenty of time and long enough for a driver to react (if necessary).

Also see **Mutual Interference** in chapter 5.5—RF Interference.

Laser Jammers
ECM—Electronic Countermeasures

Laser jammers are available, their effectiveness is not known (to the author). Most laser jammers are built into the front and/or rear license plate frame, or attach to the license plate mounts. Some laser jammers transmit all the time, some only when a signal is detected.

Some systems that only transmit when detecting a laser radar alert the driver a laser radar has been detected, some systems do not. Because laser radars have such narrow beams a laser radar at close range (by not illuminating the target license plate) could easily track a target without the detector ever alerting the system or driver.

The laser jammer beam must be relatively wide (10's of degrees) compared to the laser radar (about 0.2°); the wider the beam the more area covered and the more power required. If the jammer depends on masking (instead of fooling) valid target reflections (pulses), even more power is required. Many laser radars are designed to detect this type of brute force jamming and alert the operator.

Donald S. Sawicki

Chapter 7.2

The Courtroom

Outline
- References
- Radar Hardware Issues
  —FCC License
  —Laser Radar Use
  —NHTSA / IACP Consumer Products List (CPL)
- Manual on Uniform Traffic Control Devices (MUTCD)
- Take it to Court?
- Preparation for Court
- Court Day
- The Cosine Effect Defense

CAVEAT: This document is not intended to give legal advice in any way. This section and any other sections or references to the law or courtroom activities is based solely on the author's personal experiences and observations.

References
The prosecutor or judge will probably question ANY reference materials' accuracy and reliability, especially anything technical, presented by a defendant. Be prepared to justify the creditability of any sources or references.

All technical information and conclusions in the TRAFFIC RADAR HANDBOOK quantifiably described using illustrations, graphs, tables, or mathematical formulas—based on or derived from fundamental scientific and engineering principles, published factory specifications, measured data, or U.S. Government documents.

Radar Hardware Issues

FCC License
The United States Federal Communications Commission (FCC) classifies microwave traffic radars under Radiolocation Service. The FCC specifics technical standards such as operating frequency, bandwidth, power density, etc.

Traffic Radar Handbook
A Comprehensive Guide to Speed Measuring Systems

The FCC Rules do NOT cover the CALIBRATION of radar units, radar ACCURACY, or OPERATOR capability requirements.

LICENSE REQUIREMENTS

- State or local government agencies (including police) that have an FCC license for a communication system (Public Safety Radio Services) are not required to have a separate FCC license for traffic radar under part 90 of FCC rules.
- Radar units may also be used under Part 90 (other appropriate FCC radio license required) by non-public safety entities such as professional baseball teams, tennis clubs, automobile and boat racing organizations, private transportation firms, railroads, etc., to measure the speed of objects or vehicles.
- Many public safety agencies operate unattended, low-power, transmit-only radar units under Part 15 of FCC Rules.
- Non-public safety users are required to obtain a Part 90 license.

from FCC Public Notice, DA 96-2040, 1996 DEC 9

SUMMARY: Police do not need an FCC license to operate traffic radar (if their radios are licensed); however other state, local, or agency requirements may apply.

Laser Radar Use

Some states limit or restrict laser radar use. A New Jersey Appeals Court upheld (in a 2-1 decision) a lower court ruling that leaves in place restrictions on laser radar use. New Jersey Troopers can only use laser radar in clear weather and for targets less than 1000 feet. Source: National Motorists Association NEWS, Jan/Feb 2000, vol 11, issue 1.

NHTSA / IACP Consumer Product List (CPL)

The National Highway Traffic Safety Administration (NHTSA), in conjunction with the International Association of Chiefs of Police (**IACP**) and the Law Enforcement Standards Laboratory of the National Institute of Standards and Technology (NIST) developed testing protocols and performance standards for speed measuring devices. Radars (microwave and laser) that meet the standards are included in the NHTSA / IACP Consumer Product List (CPL) of approved speed measuring devices. ONLY SINGLE ANTENNA

CAPABILITIES (not dual antennae, if any) are tested (approved); approval not based on any time/distance measurement features (not tested and not approved).

CPL approval insures the radar (microwave or laser) meets basic minimum standards (at least when radar is factory fresh) set by the NHTSA, IACP, and NIST. Some, not all, states and/or agencies require any/all speed measuring devices (under the state or agency jurisdiction) meet CPL standards.

Manual on Uniform Traffic Control Devices (MUTCD)

The Manual on Uniform Traffic Control Devices (MUTCD), established by federal law in 1966, sets national standards for every sign, signal, pavement marking, and traffic control signal device in the USA. The standard is intended to insure signs and signals LOOK and are **USED** in the same manner everywhere. Speed limits (USE of speed limit signs) are to be determined by an Engineering Study (as defined by MUTCD 1A.13). An Engineering Study must also be done (and documented) before any speed limit can be changed; if some government body changes a speed limit without a proper study the speed limit is illegal.

Take it to Court?

Police know the uniform, rollers (flashing police lights), siren, car markings, loaded gun(s), and the power to arrest are intimidating factors when issuing a speeding ticket. Some officers are intentionally intimidating to help insure tickets are not challenged in court. Don't let intimidation alone be a factor in determining whether to challenge a ticket. Do keep in mind the odds of successfully beating a radar ticket (innocent or not) are against the defendant, but not zero.

Several factors must be considered before deciding whether to fight a speeding charge in traffic court. First and most important, one must be wrongly accursed—do not try to beat a charge if you're guilty. However, if you were not speeding as fast as accused, you might plead for the lesser speed violation to lower the fine (if you can convince the judge). With some insurance companies the higher the speed the more the insurance rate increases on conviction. The court clerk should know if the fine would change for a lower speeding ticket, and your insurance agent should know if rates change with speed.

A lawyer should be considered if one absolutely cannot afford to lose because of previous tickets and/or the effect on insurance rate. The defendant is outnumbered in the courtroom—there are one or more police officers and the prosecuting attorney(s) versus you—a lawyer can help even out the odds. However, legal representation may be more expensive then the fine, and by no means a guarantee of winning.

In general the more people the court serves the better chance for a fair hearing. A small city or county court is usually the hardest to prove a case and in some situations it is virtually impossible to walk away without paying a fine. In Democrat controlled Madison County, Illinois, for example, a court cost can be imposed on a defendant even when the case is dismissed—you pay Madison County (DEMOCRATS) for their mistake (INCOMPETENCE compounded by THIEVERY), not to mention waste your time. Some courts require the defendant to plead guilty or not guilty in a pre-trial hearing at which time a trial date is set for a not guilty plea. This means two court appearances. Some courts let defendants mail in the plea so only one court appearance is necessary.

Preparation for Court

If you go to court, be prepared to document as many facts as possible. Detailed maps could be useful—show (measure) distances, speed limits, traffic conditions, hills, valleys, obstructions, trees, parked cars, and anything else that might apply. Show your vehicle as well as the police car / radar location, but do not clutter maps with useless information. Detailed maps may sometimes be found at local or university libraries; for many areas U.S. Geological Survey maps are available. A number of Internet sites make available satellite photographs for most of the planet (see Web Links To Other Sites page). Be careful the maps or photographs are not outdated. Photos or a video tape (make sure the court has a VCR the defendant can use) of the scene can also add creditability to your case.

Some experts suggest a defendant should file a "Discovery motion" or a "request for Discovery form" with the court clerk for some or all of the following items;

- FCC Public Safety Radio Services, or Radar, **license**.
- **Radar** documents;
    —make, model, serial number, options, age,
    —manufacturer certificate of calibration,
    —operator manual and specifications,
    —calibration log sheets.
- **Tuning Fork** documents;
    —specifications (X, K, Ka; speed in mph/kmh; resonance tolerance in Hz),
    —calibration log sheets.
- Officer **Training / Qualifications**;
    —certificate of competency,
    —training material,

Donald S. Sawicki

    —officer training records.
- Departmental, local, and state **Policies** pertaining to radar use.
- officer **notes**.

Note: FCC license and Tuning Forks are not applicable to laser radar (lidar or ladar).

You or your lawyer must go through the court clerk to request discovery information by filling out a form, or making a formal request in writing regarding specific information. In most, if not all, cases an additional processing/copy fee is required. If the officer cannot produce a reasonable request for information, the charge may (should) be dismissed.

Some schools of thought suggest that requesting information might negate any chance of the officer not showing up for court (in which case the charge should be dismissed). Also, if a defendant does not make proper use of requested information the defendant's creditability could suffer.

Below is a general list of information and questions that could apply to a case in traffic court. Not all questions or details are appropriate for all cases. The better you are prepared, the better the odds of a favorable outcome (the same applies to the prosecution, except the odds are already in their favor). The more pertinent questions you can ask that the officer cannot answer, the stronger your case.

Police Officer
Typical Court Presentation

Most officers are trained to collect certain information in the event the case is disputed. Below list typical minimum information an officer should be prepared to present in court.

- Establish time, place, location of radar.
- Establish location of offending vehicle.
- Identify (able to) the offending vehicle.
- Establish and identify the vehicle operator.
- Visually observed apparent excessive speed.
- Observed vehicle was out front, by itself, nearest the radar during readings (not Fastest Target mode).
- Established steady stable target track history.
- Insured minimum interference from external sources.
- Establish RADAR tested before (and/or after) use;
  —self-test run,
  —tested with tuning fork (not laser radar),

—tested for Range Accuracy (laser radar),
—Scope / Beam alignment tested (laser radar),
—and/or tested against test vehicle (with calibrated speedometer).
• State qualifications and training.

POSSIBLE QUESTIONS
the officer might be asked

Radar Accuracy Verification

• Is the radar certified and who established guidelines for certification?
• Who established and what are policy guidelines for radar **operation**?

    Was the site pre-tested for radar operation?
    (run a test target at different speeds to verify radar accuracy)
    —Speeds test target run at?
    —Max and min distances?
    —Any Cosine Effects (large angles only) observed? (chapter 4)
    Any warm-up time required on initial power-up?
• Who established and what are policy guidelines for radar **testing**?
    Tuning fork test done? (chapter 5.2)
    (Tuning forks do not work with laser radars)
    Moving mode radar requires 2 tuning forks to test moving mode
    —how many forks? (speed, radar band or frequency label/sticker?)
    —tuning fork calibration data maintained / available?
    Radar self-test? (chapter 5.2)
    How often is self-test run?
    —automatic (on power-up, periodically, before target measured) and/or operator initiated?
    —test / adjust transmit frequency?
    —check supply voltage (car or internal battery)?
    —test target signal?
    —test portions of the digital circuits?
    —test / adjust portions of the analog circuits?
    —test display indicators?

    Laser Radars (Ladars or Lidars) (also see chapter 6.5)
    Range accuracy test done? At what range? (chapter 6.6)
    Alignment (aim sight-to-apertures) test done?
    At what range? (chapter 6.4)

(alignment test should be done at max operating range for both horizontal and vertical axis)
- Who established and what are policy guidelines for radar calibration?
  Date radar last calibrated?
  Period between calibrations?
  Calibration log sheets available?
  —Date, tests, adjustments, conditions, etc.
  Who calibrates radars?

Alleged Violation

- Check officer's memory and/or **notes** of alleged violation.
  —Time, Place, Location of radar?
  —Driver description
  —Ticketed Vehicle description? (note any custom features)
  —Number of tickets officer issued that day? that month? this month?

- **Target mis-identified**
  Traffic? (heavy, medium, light, none)
  —any other targets in beam?
  —radar beamwidth?
  —lanes covered at target distance?
  —typical detection range? (small, medium, large targets)

- **Track History**
The longer the radar tracks a target the better chance for an accurate reading, and the less chance of **interference**.
  —range / time radar first tracked target?
  —range / time of last track?
  —total distance / time radar tracked target?
Some department require a minimum of several seconds to establish track history

- **Possible Interference Sources**
  Does radar have any RFI (Radio Frequency Interference) indication?
  Any near-by transmitters?
  —other traffic radars?
  —patrol car radios and/or walkie-talkies?
  —near-by patrol car(s) transmitters?
  —CBs, HAMS, other?
  Heater/Air-conditioning and/or fan operating?

        Engine running?
- Radar distance off the road
or distance between target-patrol lanes (moving radar)?
        _____ ft or m
        —Note Cosine Effect angle vs measured speed
- Did operator or electronic circuits (Auto-lock) lock target?
(some courts will not accept radar evidence if Auto-lock featured used)
- Moving mode radar
    —Patrol car speed?
    —Speedometer match radar patrol car speed?
    —shadowing? (chapter 5.4)
    —batching or speed bumping? (chapter 5.4)
    —Antenna alignment (_____degrees) to patrol car direction?

Radar Hardware
Establish officer's knowledge (or lack) of equipment.

- Radar MAKE and MODEL, Options?
- Configuration
    —single unit (chapter 5.4—Scanning Error)
        - handheld (location?)
        - mounted (location?)
    —multi-piece? (chapter 5.4—Panning Error)
        - forward antenna (location?)
        - rear antenna (location?)
        - handheld antenna (located?)
- stationary—moving-mode? (switch properly set?)
- multiple antennas? (switch properly set?)
    —which antenna(s) active?
- Target Direction(s)? (switch properly set?)
    —stationary—on-coming / going?
    moving (front antenna)—opposite direction / same-lane traffic?
    moving (rear antenna)—opposite direction / same-lane traffic?
    moving (front and rear antenna)—opposite direction and/or same lane (front and /or rear)
- RFI (Radio Frequency Interference) Indicator?
    light/LED/message—audio alert tone (speaker)—both
- speed measuring range
    —target min and max (stationary mode)?
    —target min and max (moving mode opposite direction target)?
    —patrol car min and max (moving mode opposite direction target)?

Donald S. Sawicki

   —patrol car min and max (moving mode same-lane target)?
   —target/radar min speed difference, target and radar min speed? (same-lane target)?
 • accuracy (entire speed measuring range checked?)
   —stationary
   —moving mode (target)
   —moving mode (patrol car)
 • Radar band? (X, K, Ka, laser)
 —if Ka band, fix frequency or hopper?
     if Ka fixed frequency, what frequency? _____GHz
     if Ka hopper, frequencies? _____,_____,_____,_____,_____,_____ GHz
 • Integration (sample) Time _____ms or sec
 radar display update time _____updates / _____ sec or ms
 samples / display-update _____
 • Beamwidth? _____ horizontal (degrees) _____ vertical (degrees)
 • auto-lock feature (photo radar and some older models)?
 • maximum detection range? (small target—Medium target—LARGE target)
 • Range (threshold) setting? (min—in-between—max)
 • operating temperature range?
 • date radar manufactured?
     year purchased?
     new or used?
     if used was it refurbished by factory?
 • numbers and types of radars in department?

Training and Qualifications
 • Hours of formal training?
 —who trained (and trainer qualifications)?
 —certificate of competency?
 —training material?
 —training records?

Court Day

When the court date finally comes be prepared to spend the entire day. Do not be surprised that typically 100 or more people show up at the same time. One of the prosecutors may try to plea bargain the case before trial with an offer that depends on how strong he surmises the case is after questioning you. Do not tip too much information in the event the case goes to trial. Be prepared; scare tactics, intimidation and time-consuming distractions are designed to make you believe that a deal is better than going before the judge. Do not be maneuvered into a bad deal; many traffic court prosecutors are amateur beginners. If possible, listen to what other defendants are bargaining for. Bargaining to pay the fine and to not have the ticket on record (for the insurance company to see) is sometimes the best deal; a short probation period of 30 to 90 days may also be required (otherwise the ticket goes on record).

If you do go before the judge do not try to persuade the prosecutor any longer, the judge is the only one to be concerned with at this point. Don't try to make a case with material you don't thoroughly understand; most judges recognize a shaky case. Do not be afraid to ask the officer who was operating the radar specific, pertinent questions about the radar. The more questions the officer cannot answer, the less creditable the prosecution's case. If a defendant can establish the officer's lack of working knowledge about radar the defendant's chance of winning may increase.

The general format for traffic court is the prosecution will state their case and question the officer for support. You will then have a chance to state your case and question the officer. The prosecutor will then try to shatter your case by asking you questions. Be aware judges have authority to conduct court procedures to their satisfaction; this means that court procedure varies with state, county, city, and judge or magistrate (Federal).

The Cosine Effect Defense

Just because the Cosine Effect (Chapter 4.1) works in favor of the motorist most of the time (moving mode radar has an exception [chapter 4.3]) does not mean one cannot become a victim of the Cosine Effect.

If an officer is operating a laser or microwave radar far off the road, the Cosine Effect could be so severe the radar measures a relatively small percentage of actual target speed. It is quite possible the corrected target speed calculates excessively high, which indicates the radar was tracking some other target or signal. For example if the cosine angle is 60 degrees radar measured speed is half true target speed. If the officer states observed estimated target speed to be about the same as the radar measured speed (half true speed), one could argue the officer's judgment was biased incorrectly by the radar. If the officer's estimated target speed conflicts with radar corrected (for Cosine Effect) speed, which one is correct, if any?

Another Cosine Effect limitation concerns radar, microwave or laser, minimum range—targets inside minimum range are not measured (because relative speed is changing too fast). A target inside minimum range could be the closest target, but if another target is outside minimum range but close enough for the radar to measure, a radar operator could mistake the radar reading as the close (inside minimum range) target instead of the more distant target.

Donald S. Sawicki

The greater off target path (lane) the greater the Cosine Effect and the greater radar minimum range. Radar sample time, speed tolerance (accuracy), and target speed are also factors, but distance off target lane has the greatest influence. See chapter 4.2—Cosine Effect on Measured Speed / Minimum Range.

For example a radar 90 feet (30 yards) off target lane with a 300 ms sample period and ±1 mph accuracy cannot measure targets (coming or going) traveling at 65 mph when target range is 230 feet (77 yards) or less. The greater the speed the greater the minimum range; minimum range for targets traveling at 85 mph is 281 feet which equals 94 yards (17 yards greater than a 65 mph target). Minimum range numbers derived using the Compute Range Given Acceleration calculator.

The example image location is Interstate 55 (replaced Route 66) south of Springfield, Illinois (USA) at the truck Weigh Station. State police have been known to operate (as shown) laser and microwave radar from this station and the station north of Springfield (virtually identical to the southern station except for south bound traffic). At both stations both directions of travel are 3 lanes, distance between (medium) north and south bound traffic is approximately 25 yards.

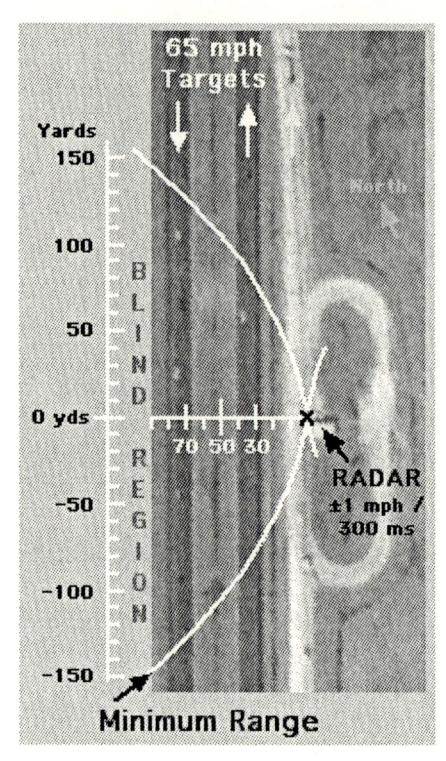

Traffic Radar Handbook
A Comprehensive Guide to Speed Measuring Systems

Chapter 8.1

RF Biological Effects

Outline
- Radiation
- Studies and Reports
- Electrical Properties of Living Matter
- Thermal Effects of RF Radiation
- Interaction of Fields and Biological Systems

Radiation

Natural radiation was the only source of human exposure until the latter part of the nineteenth century when Thomas Edison invented the electric light. Most natural radiation of significance occurs in a small part of the lowermost frequency spectrum (electrostatic or less than about 50 Hz) and in the uppermost part of the spectrum (above 10 THz or 10^{13} Hz); man-made radiation dominates 50 Hz to 300 GHz, and accounts for some radiation in the lower and upper spectrum.

Most natural radiation below 5 kHz results from lightning. Average rate of global lightning strikes is about 100 bolts per second (cloud to ground and cloud to cloud). Some natural radiation below 5 kHz results from pulsations in the earth's magnetosphere during intense solar storms—such as when the Aurora over the poles (north and/or south) is active.

Radiation is categorized as either non-ionizing or ionizing radiation. Ionizing radiation has enough energy (high enough in frequency) to break atomic bonds by removing one or more electrons (ionization is adding or removing electrons) and creating a charged atomic particle; non-ionizing radiation does not. The higher the frequency (the lower the wavelength) the greater the energy and ionization.

Donald S. Sawicki

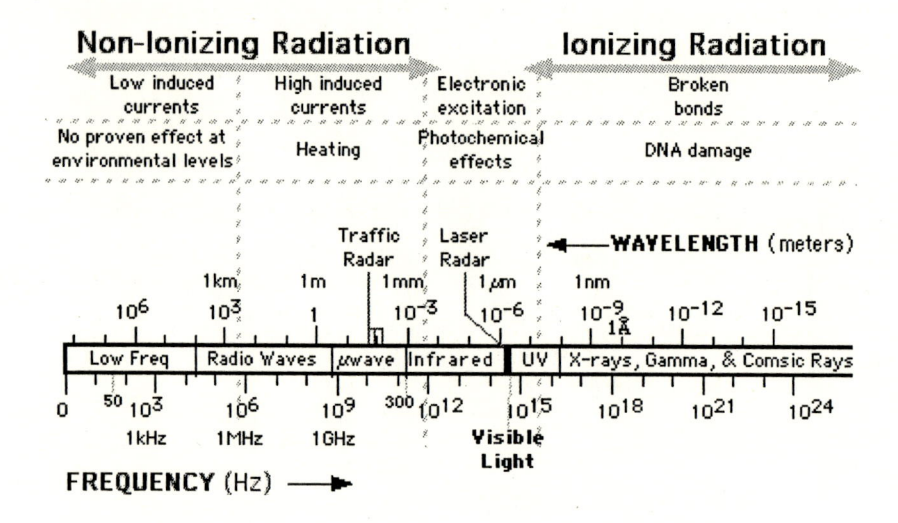

Figure 8.1-1—**Non-Ionizing / Ionizing Radiation**

Non-ionizing electromagnetic radiation is generally divided into three categories; electrostatic (non time-varying), low frequency (such as house current), and RF or radio frequency (greater than about 30 kHz and less than 300 GHz). Other forms of radiation electromagnetic in nature are in the uppermost part of the frequency spectrum (above 10 THz or 10^{13} Hz) and include IR—infrared (heat), visible light, and ionizing radiation such as ultraviolet (UV), X-rays, gamma rays and cosmic rays. All forms of radiation can have adverse health effects when intense enough and/or time exposure long enough.

Studies and Reports

Microwave traffic radar has been linked to several adverse health effects by a number of police officers who operated traffic radars (low level microwave radiation) over a long period (hours at a time for years). To date the scientific community cannot establish a mechanism that links traffic radar to adverse health effects (that does not mean one does not exist). Data on traffic radar and health is pitifully scarce.

The U.S. Air Force sponsored a study of rats exposed to pulsed microwaves at 24.5 GHz. Note that K band traffic radars operate at around 24.025 - 24.250 GHz (475 - 250 MHz lower); also the study used pulsed microwaves and traffic radar is continuous (not pulsed). The research cost 54 million dollars and was conducted by the University of Washington School of Medicine in Seattle and published in 1984. The study showed a significant increase in malignant tumors and noted affects in the adrenal glands and the entire endocrine system. This investigation suggests the maximum allowable exposure for humans set by the ANSI (American National Standards Institute) and IEEE (Institute of Electrical and Electronic Engineers) are too low; human exposure limits should be more stringent.

The London Times reported on 1998 December 31 that Dr. Henry Lai, an expert in non-ionizing radiation and professor at the School of Medicine and College of Engineering at the University of Washington, Seattle, announced that low-level microwave radiation can split DNA molecules in the brains of laboratory mice. DNA is Deoxyribonucleic acid; a complex, usually helical shaped chemical compound that is the substance that makes the organic matter of genes and chromosomes. Splitting DNA molecules in the brain is associated with Alzheimer's and Parkinson's Disease, and cancer. The cellular telephone industry supported Dr. Lai's research grant, but suppressed the report's publication.

Electronic Engineering Times reported 2001 June 11 the Air Force Research Laboratory (AFRL) has developed a non lethal antipersonnel millimeter (mm) band "heat ray" intended for use on battle fields or against hostile crowds. A 3 x 3 meter (about 10 x 10 feet) antenna (mounted on a Humvee, aircraft, helicopter, or ship) can be swept across a crowd with a 95 GHz beam that induces heating in the skin. Reports indicate a 2 second burst can heat the skin to 130°F—inflicting pain with an intense burning sensation. Officials claim the energy only penetrates the top 1/64 inch (0.4 mm) of the skin and is not harmful to internal organs (they have not proved to me the eyes are not vulnerable to damage). Operating range is believed to be around 700 yards (640 meters) to 1090 yards (1000 meters); rain, sea spray, fog, and humidity will reduce range. Possible countermeasures include shielding the energy using very thick clothing, a metallic sheet (aluminum foil), or a metal trash can lid.

Many studies on health effects and exposure to electromagnetic fields conducted since 1948 have reached varying and sometimes contradictory conclusions. Much of the early research concentrated on the thermal heating affects of electromagnetic radiation; some later studies indicated reactions to EMF exposure not explained by thermal heating. Experts do not always agree on the levels or types of electromagnetic fields that affect health.

Electrical Properties of Living Matter

Living matter exhibits many electric properties as well as generates various, relatively small, electromagnetic fields. Medical doctors use known and well-documented electrical properties of the body to determine health and diagnose

Donald S. Sawicki

problems. This section is a brief sample of some electrical properties of living matter.

- Nerve fibers consist of cylindrical membranes with one conducting fluid inside another conducting fluid and a potential difference of about 0.1 volts between the fluids. A pulse causes the membrane between the fluids to temporarily become more permeable to ions and the voltage drops. A pulse travels approximately 98 feet per second (30 m/s) or about 67 miles per hour (108 kmh).
- Mechanical energy from bone bending and stress creates weak electrical potentials (voltages) of a few millivolts across a centimeter and relatively low frequency.
- Electrocardiographs (EKGs) measure potential differences (voltages) between the chest and back to study heart functions. The human heart also has an electric field near the surface (of the heart) that is between 1 and 10 volts per meter (V/m).
- Electroencephalographs (EEGs) measure potential differences, on the order of microvolts, in the scalp and are a meager measures of brain functions. EEG patterns (brain waves) are different for every person, similar in twins, and similar for certain brain disorders such as epilepsy, brain tumors, brain damage, encephalitis, systemic diseases (toxemia and diseases of liver and kidneys). When resting but not asleep, the back part of the head will register alpha waves (or alpha rhythms) that have a frequency of 8 to 12 hertz. Beta waves, 18 to 25 hertz, relate to sensory functions and are smaller in magnitude than alpha waves. People in comas have patterns with 1 to 3 hertz rates near the damaged area of the brain. Theta waves are 4 to 7 hertz—normal in infants and young children but abnormal in adults.

Thermal Effects of RF Radiation

Electromagnetic energy is well known to cause thermal heating in living tissue. Microwave ovens use electromagnetic energy to heat and cook food. A microwave oven is just a magnetron oscillator (radar transmitter) operating at 2.45 GHz. The amount of heating that takes place is a function of transmit power and duty cycle (time). Maximum surface heating due to RF exposure of a typical human occurs at frequencies of around 30 to 120 MHz.

Tissue heating depends on the frequency of the source and the dielectric constant, water content and thickness of the tissue. The more conductive the tissue the more energy absorbed and heat generated. It requires a relatively large amount of radiation to heat tissue. Radiation levels too low to produce heat may have other effects at a cellular level, although not all experts accept this. Some experts believe most non thermal health effects require much higher field levels compared to thermal heating effects.

Fields strong enough to cause heating require hundreds to thousands of watts. Localized heat of about 1 watt per kilogram (0.45 W/lb) can damage tumors. The temperature of the tumor is raised to between 43 and 45 degrees Celsius (109 and 113 degrees Fahrenheit). The FCC allocates frequency bands for medical use that include 13.56, 27.12, 40.68, 915, and 2,450 MHz (microwave ovens operate at 2.45 GHz); other frequencies have shown better results in some cases. Fields that cause mild heating can promote tissue healing or relax muscles.

Interaction of Fields and Biological Systems

Weak RF fields insufficient to cause heating but strong enough to induce peak potentials of 1 to 1.5 millivolts per centimeter (0.10 to 0.15 V/m) can promote healing of broken bones. Experiments have shown osteoporosis, the loss of bone mass, can be halted or reversed by pulsed RF radiation. The shape and timing of pulses is extremely significant, and different, to promote bone healing or affect osteoporosis.

Some studies indicate there are reactions to RF exposure not explained by thermal heating. Strong RF fields heat tissue by vibrating molecules of the tissue. Weaker fields can induce electrical currents in or on tissue, the stronger the field the larger the induced current. Electric and magnetic fields can produce weak Lorentz forces that may affect charged particles (ions) on a molecular scale. The Lorentz force is the force on a charged particle in motion due to the presence of an electric and magnetic field, and may boost or inhibit cell chemistries by pumping ions.

Listed below is a sample of some observed affects of electromagnetic radiation on living cells. Most if not all effects depend greatly on frequency, modulation, and magnitude of the field.

- split DNA molecules in the brains of laboratory mice
- Peral chain—randomly suspended particles (such as fat globules and E. coli bacteria) align with each other in direction of field.
- Non spherical particles (such as E. coli) line up either perpendicular or parallel to electric field depending on frequency.
- Particle movement.
- Change in natural shape of cells.
- Cell death from membrane damage.
- Fusion of cells.

Donald S. Sawicki

RF radiation has been linked to biochemical effects, immunological effects, Alzheimer's and Parkinson's Disease, cancer, cataracts, EEG effects and behavioral changes, to name a few. Studies from the old Soviet military suggest that some frequencies and modulations cause behavioral changes in humans. In 1990 the U.S. military was reported to be planning microwave radiation experiments on animals to study behavioral effects.

Some health effects seem to take place in small bands or windows of frequencies, modulations, and magnitudes. For example, nerve tissues affected by continuous 60 hertz fields are unaffected by 55 or 65 hertz fields. Some experiments have shown for some narrow frequency bands and specific modulation types; a smaller field affects cells more than a stronger field. Several factors should be considered to determine electromagnetic radiation exposure and include;

Field Type—Frequency and modulation; modulation introduces other frequency components.
Distance from Source—Field magnitudes change greatly with distance.
Exposure Time—Exposed to large fields for a short period of time or exposed to small fields for a long period of time.

Traffic Radar Handbook
A Comprehensive Guide to Speed Measuring Systems

Chapter 8.2

RF Radiation Standards

Outline
- American National Standards Institute (ANSI)
- Institute of Electrical and Electronic Engineers (IEEE)

The Federal Communications Commission (FCC), the Environmental Protection Agency (EPA), the Occupational Safety and Health Administration (OSHA), and the National Institute for Occupational Safety and Health (OIOSH) are federal agencies that have established standards regarding radio frequencies. The American National Standard Institute (ANSI) and the Institute of Electrical and Electronic Engineers (IEEE) are private organizations that have well-established standards regarding radio frequencies. Some state and local governments have also established limits for radio frequency and/or 60 Hz emissions. Depending on the device, intended use and place used (commercial, industrial, business, private, city/state) determines which standards apply—for most situations, if not all, more than one standard applies. This section considers ANSI and IEEE safety limits with respect to human exposure to radio frequencies for the general public.

American National Standards Institute (ANSI)
American National Standards Institute (ANSI) standard C95.1-1982 sets electric and magnetic field (see Appendix B—Electromagnetic Waves) strength limits for the general public for frequencies between 300 kHz and 100 GHz. Below 300 MHz the electric and magnetic fields must be accounted for separately. The below figure illustrates the electric field strength (E field) limits in volts per meter (V/m), the magnetic field strength (H field) limits in amperes per meter (A/m) and where applicable the power density limits in milliwatts per square centimeter. The limits are based on whole body exposure averaged over 0.1 hours (6 minutes); an absorption rate of 0.4 watts per kilogram (0.18 W/lb) is considered a safe limit.

Donald S. Sawicki

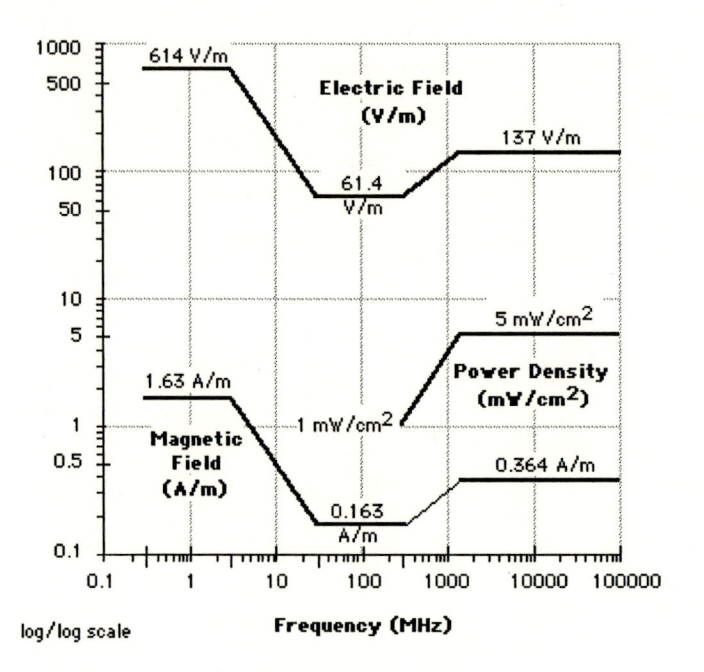

Figure 8.2-1—ANSI Standard

At traffic radar frequencies (X, K, and Ka) the ANSI limit is 5 milliwatts/square centimeter. This equates to an electric field strength of 137 volts/meter and a magnetic field strength of 0.364 amperes/meter.

Institute of Electrical and Electronic Engineers (IEEE)

The Institute of Electrical and Electronic Engineers (IEEE) standard C95.1-1991 sets electric and magnetic field (see Appendix B—Electromagnetic Waves) strength limits for the general public for frequencies between 3 kHz and 300 GHz. Below 100 MHz the electric and magnetic fields must be accounted for separately. The below figure illustrates the electric field strength (E field) limits in volts per meter (V/m), the magnetic field strength (H field) limits in amperes per meter (A/m) and where applicable the power density limits in milliwatts per square centimeter.

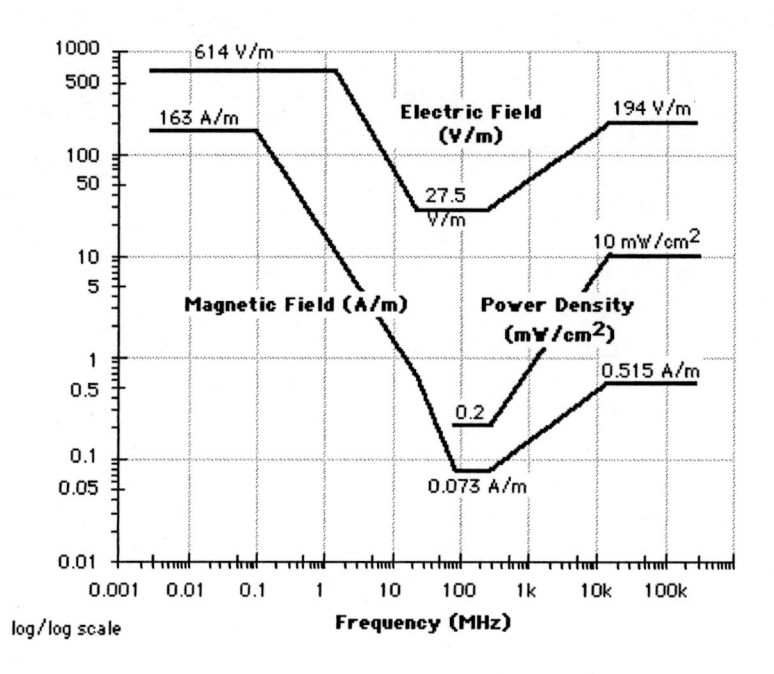

Traffic Radar Handbook
A Comprehensive Guide to Speed Measuring Systems

Figure 8.2-2—**IEEE Standard**

At traffic radar frequencies (X, K, and Ka) the IEEE limit is about 10 milliwatts/square centimeter. This equates to an electric field strength of 194 volts/meter and a magnetic field strength of 0.515 amperes/meter.

Donald S. Sawicki

Chapter 8.3

Traffic Radar
Power Density

Outline
- Power Density vs Range given ERP
- ERP, Power Density, and Electric Field

Power Density vs Range given ERP

The below figure graphs POWER DENSITY versus RANGE for various ERPs (Effective Radiated Power). Typical X and K band traffic radar ERPs vary from about 25 to 135 milliwatts (100 milliwatts seems to be an average), based on NHTSA 1980 tests (see chapter 5.1). Ka band radars transmit slightly higher power due to atmospheric losses at those frequencies).

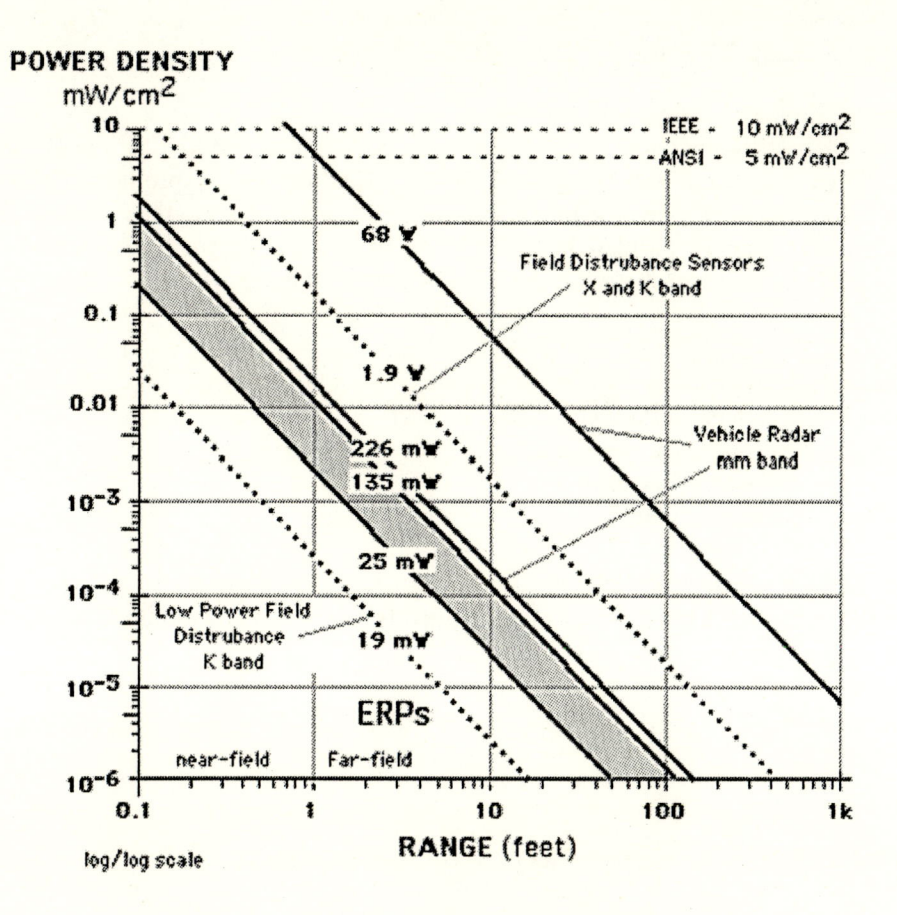

Traffic Radar Handbook
A Comprehensive Guide to Speed Measuring Systems

Figure 8.3-1—**Typical Traffic Radar Power Density**

The higher the frequency and the more oxygen and water vapor in the atmosphere, the greater the signal propagation (atmospheric absorption) loss. The above figure is for comparison purposes (to compare same or near same frequency signals) and does not account for atmospheric absorption because this varies widely with conditions (and signal frequency). Power densities in practice will be less than shown in the figure. Also see chapter 5.5—Interference, Natural Interference.

Some field disturbance sensors (alarms and automatic door openers) transmit at X and/or K band, maximum allowed electric field strength at 3 meters is 2.5 volts/meter (chapter 5.5—Interference, RF Interference). Note that 2.5 V/m at 3 m, assuming 3 meters is in antenna far-field (/Appendix E), translates to about 0.0166 watts per square meter at 3 m (1.87 watts ERP).

ERP, Power Density, and Electric Field

ERP (Effective Radiated Power) is a function of TRANSMITTER power (and cabling loss) and ANTENNA GAIN. ERP equals the product of antenna gain times power delivered to the antenna. Power delivered to the antenna equals transmitter power divided by coax cable or waveguide (and connectors) loss (signal power in/out) between the transmitter and antenna.

$$P_{ERP} = P_t\, G\, /\, L$$

P_{ERP}—Effective Radiated Power (watts)
P_t—transmitter power (watts)
G—antenna gain (ratio)
L—cable or waveguide loss (ratio signal in/out)

Power density and electric (and magnetic) field strength are a function of ERP and range from antenna. Range must be in antenna far-field (/Appendix E), typically a minimum of inches to several feet for traffic radars.

$$P_D = P_{ERP} / (4\pi R^2) = E^2 / Z$$

$$P_{ERP} = P_D(4\pi R^2) = (E^2/Z)(4\pi R^2)$$

$$E = (P_D Z)^{1/2} = \{\, P_{ERP}\, Z / (4\pi R^2)\, \}^{1/2}$$

P_D—power density (watts / meters^2)
R—range (meters)
E—electric field strength (volts / meter)
Z = 376.73 ohms (intrinsic impedance)
π = 3.14159...

Traffic Radar Handbook
A Comprehensive Guide to Speed Measuring Systems

appendix A
Frequency Spectrum

Appendix A Outline

- Introduction
    Table A-1—Frequency Multipliers

- Traffic Radar Frequencies
    Table A-2—Traffic Radar Frequency Bands
    Table A-3—Select Ka Band Traffic Radar Frequencies
    Figure A-1—Select Ka Band Radars

- Frequency Band Designations
    Table A-4—Military Radar Bands
    Table A-5—International Telecommunications Union Radar Bands
    Table A-6—Radio Frequency Bands
    Table A-7—ECM Bands

Introduction

The Federal Communications Commission (FCC) regulates frequency use by setting specifications for transmit frequency, power, bandwidth, modulation, and location, to name a few, and requires a license for many uses. The FCC does NOT cover the CALIBRATION of radar units, radar ACCURACY, or OPERATOR capability requirements.

Note the following multiplier abbreviations for frequency (also see appendix B—Electromagnetic Waves).

Hz (hertz)	cycles per second	1 Hz
kHz (kilohertz)	one thousand hertz	1,000 Hz
MHz (megahertz)	one million hertz	1,000,000 Hz
GHz (gigahertz)	one billion hertz	1,000,000,000 Hz
THz (terahertz)	one trillion hertz	1,000,000,000,000 Hz

Table A-1—**Frequency Multipliers**

Donald S. Sawicki

Traffic Radar Frequencies

BAND	FREQUENCY		NOTES
S	2.455 GHz		obsolete
X	10.525 GHz	±25 MHz	one 50 MHz channel (10.500-10.550 GHz)
Ku	13.450 GHz		no known systems
K	24.125 GHz	±100 MHz	Europe
K	24.150 GHz	±100 MHz	one 200 MHz channel (24.050-24.250 GHz)
Ka	33.4-36 GHz		13 channels; 200 MHz/ch
IR	333 THz		900 nm wavelength Infrared (IR)

Table A-2—**Traffic Radar Frequency Bands**

FREQUENCY	SYSTEM
33.3 GHz	Genesis II
33.4 GHz	photo radar
33.8 GHz	BEE 36
34.3 GHz	TMT-6F photo radar
34.3 GHz	Multanova 6F photo radar
34.6 GHz	PR-100 photo radar
34.7 GHz	Stalker ATR
34.94 GHz	Stalker ATR
34.2-35.2 GHz	Stalker ATR (freq hopper)

Table A-3—**Select Ka Band Traffic Radar Frequencies**
Frequency tolerance for Ka band radars typically ±100 MHz.

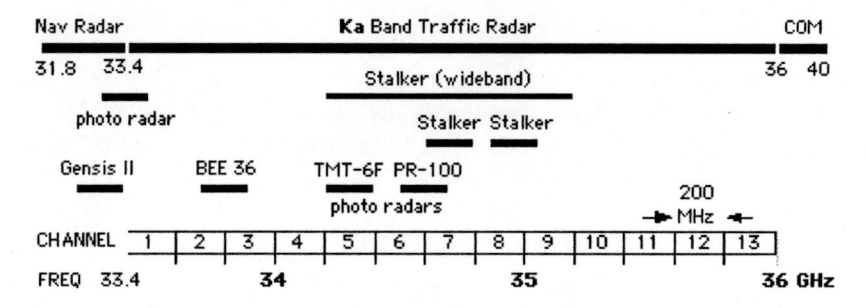

Traffic Radar Handbook
A Comprehensive Guide to Speed Measuring Systems

Figure A-1—**Select Ka Band Radars**

Frequency Band Designations

Military Radar Bands

Military radar band nomenclature (L, S, C, X, Ku, K and Ka bands) originated during World War II as a secret code so scientists and engineers could talk about frequencies without divulging them. After the war the codes were declassified, millimeter (mm) was added, and the designations were eventually adopted by the IEEE—Institute of Electric and Electronic Engineers. Military radar band nomenclature is widely used today in radar, satellite and terrestrial communications, and electronic countermeasure applications, both military and commercial.

Radar Band	Frequency	Notes
HF	3 - 30 MHz	High Frequency
VHF	30 - 300 MHz	Very High Frequency
UHF	300 - 1000 MHz	Ultra High Frequency
L	1 - 2 GHz	
S	2 - 4 GHz	
C	4 - 8 GHz	
X	8 - 12 GHz	
Ku	12 - 18 GHz	
K	18 - 27 GHz	
Ka	27 - 40 GHz	
mm	40 - 300 GHz	millimeter wavelength

Military HF, VHF, UHF same as Radio Band HF, VHF, UHF respectively.

Table A-4—**Military Radar Bands**

Donald S. Sawicki

ITU Radar Bands

The International Telecommunications Union (ITU) specifies bands designated for radar systems as described in the table below. The ITU bands are sub-bands of military designations.

ITU Band	Frequency
VHF	138 - 144 MHz
	216 - 225 MHz
UHF	420 - 450 MHz
	890 - 942 MHz
L	1.215 - 1.400 GHz
S	2.3 - 2.5 GHz
	2.7 - 3.7 GHz
C	5.250 - 5.925 GHz
X	8.500 - 10.680 GHz
Ku	13.4 - 14.0 GHz
	15.7 - 17.7 GHz
K	24.05 - 24.25 GHz
Ka	33.4 - 36.0 GHz

VHF—Very High Frequency
UHF—Ultra High Frequency

Table A-5—**International Telecommunications Union Radar Bands**

Radio Bands

Radio band designations are summarized below. Note that the radio band chart includes wavelength. In the early days of radio it was easier to measure wavelength than frequency.

Band	Nomenclature	Frequency	Wavelength
ELF	Extremely Low Frequency	3 - 30 Hz	100,000 - 10,000 km
SLF	Super Low Frequency	30 - 300 Hz	10,000 - 1,000 km
ULF	Ultra Low Frequency	300 - 3000 Hz	1,000 - 100 km
VLF	Very Low Frequency	3 - 30 kHz	100 - 10 km
LF	Low Frequency	30 - 300 kHz	10 - 1 km
MF	Medium Frequency	300 - 3000 kHz	1 km - 100 m
HF	High Frequency	3 - 30 MHz	100 - 10 m
VHF	Very High Frequency	30 - 300 MHz	10 - 1 m

Traffic Radar Handbook
A Comprehensive Guide to Speed Measuring Systems

UHF	Ultra High Frequency	300 - 3000 MHz	1 m - 10 cm
SHF	Super High Frequency	3 - 30 GHz	10 - 1 cm
EHF	Extremely High Frequency	30 - 300 GHz	1 cm - 1 mm

Table A-6—**Radio Frequency Bands**

ECM Bands

The electronic countermeasures (ECM) industry occasionally refers to band designations as described below.

Band	Frequency
A	30 - 250 MHz
B	250 - 500 MHz
C	500 - 1,000 MHz
D	1 - 2 GHz
E	2 - 3 GHz
F	3 - 4 GHz
G	4 - 6 GHz
H	6 - 8 GHz
I	8 - 10 GHz
J	10 - 20 GHz
K	20 - 40 GHz
L	40 - 60 GHz
M	60 - 100 GHz

Table A-7—**ECM Bands**

Donald S. Sawicki

Traffic Radar Handbook
A Comprehensive Guide to Speed Measuring Systems

appendix B
Electromagnetic Waves

Appendix B Outline
- Electromagnetic Wave Parameters
- Electromagnetic Wave Propagation
- Polarization

Electromagnetic Wave Parameters

The most common and useful parameters for describing electromagnetic waves are frequency, wavelength and period. The figure below represents an electromagnetic wave in a time and a distance coordinate system.

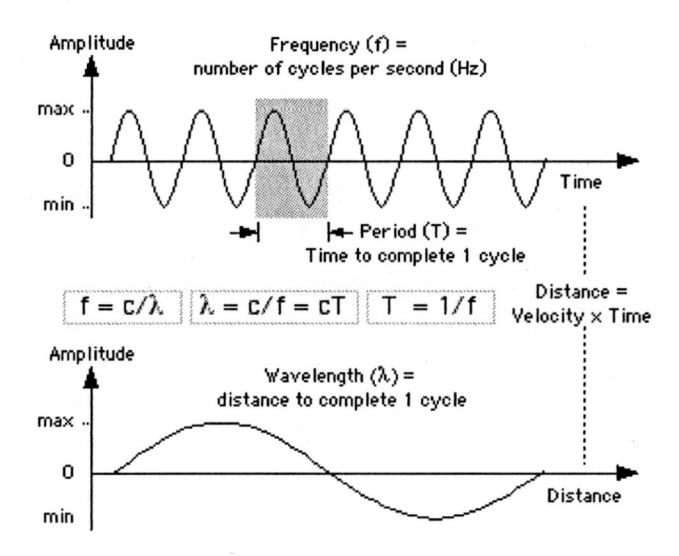

$$f = c/\lambda \quad \lambda = c/f = cT \quad T = 1/f$$

Figure B-1—**Sine Wave Parameters**

c = speed of light
299,792,456.2 meters/second (vacuum)
186,282 miles/second
670,616,625 miles/hour (mph)
983,571,051 feet/second
approximately 1 foot/nanosecond

Radar Band	Frequency (GHz)	Period (nanoseconds)	Wavelength (inches)	Wavelength (meters)
S	2.455	0.407	4.8	0.122
X	10.525	0.095	1.12	0.0285
K	24.15	0.041	0.489	0.0124
Ka	33.4 - 36.0	0.030 - 0.028	0.353 - 0.328	0.00898 - 0.00832
laser (IR)	333,100	0.000003	0.000035	900 nanometers (nm)

Table B-1—**Traffic Radar Wavelengths**

Source	Frequency	Period	Wavelength	
Europe AC	50 Hz	0.0200 sec	3,726 mi	5,996 km
US AC	60 Hz	0.0167 sec	3,105 mi	4,997 km
AM Radio	530 - 1700 kHz	1.89 - 0.589 μs	1856 - 579 ft	566 - 176 m
Citizens Band	26.96- 27.41 MHz	37.09- 36.48 ns	36.48-, 35.88 ft	11.12- 10.94 m
Cordless Phones	46.6- 49.9 MHz	21.46- 20.04 ns	21.11- 19.71 ft	6.433- 6 m
FM Radio	88- 108 MHz	11.36- 9.26 ns	11.18- 9.11 ft	3.407- 2.776 m
Police Radio (typical)	155 MHz	6.45 ns	6.35 ft	1.93 m
UHF TV	470- 806 MHz	2.13- 1.24 ns	25.11- 14.64 in	0.6379 - 0.372 m
Cellular Phones	824- 894 MHz	1.21- 1.11 ns	14.32- 13.2 in	0.364- 0.335 m

Table B-2—**Wavelengths for Common Sources**

Electromagnetic Wave Propagation

Radio waves, radar waves, microwave oven waves, and light waves all have the properties shown in figure B-2. These properties consist of an electric field and a magnetic field at right angles to each other. The electric field (E field) is measured in volts/meter and the magnetic field (H field) is measured in amperes/meter; this is analogous to voltage and current phenomena in circuit analysis. The ratio of electric field over magnetic field in free space is 377 ohms. Power is a function (vector cross product) of the electric and the magnetic fields.

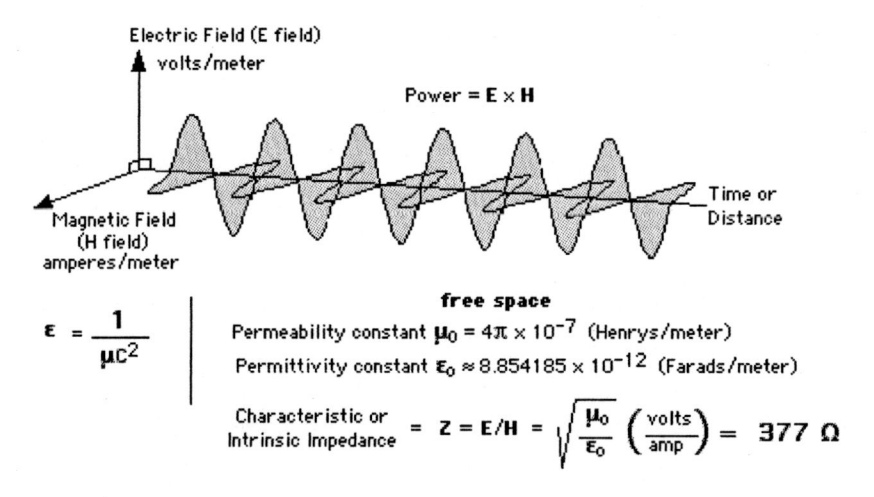

$$\varepsilon = \frac{1}{\mu c^2}$$

free space
Permeability constant $\mu_0 = 4\pi \times 10^{-7}$ (Henrys/meter)
Permittivity constant $\varepsilon_0 \approx 8.854185 \times 10^{-12}$ (Farads/meter)

$$\text{Characteristic or Intrinsic Impedance} = Z = E/H = \sqrt{\frac{\mu_0}{\varepsilon_0}} \left(\frac{\text{volts}}{\text{amp}}\right) = 377\ \Omega$$

Figure B-2—**Electromagnetic Wave in Free Space**

Polarization

Polarization is defined as the plane of the electric field. For example, in the U.S., television waves are horizontally polarized while commercial AM and FM broadcast are vertically polarized (as in above figure). Radar waves may be polarized horizontally, vertically or in between. Circular polarization is another technique used to propagate electromagnetic waves. With circular polarization the plane of the electric field is not constant but rotates at a rate proportional to frequency. Circular polarization tends to propagate through rain better than other polarizations.

Donald S. Sawicki

Traffic Radar Handbook
A Comprehensive Guide to Speed Measuring Systems

appendix C
Doppler Equations

Basic Equation for Doppler Shift

A stationary radar will transmit a signal and a moving vehicle (radar detector in vehicle) will receive a signal Doppler shifted by the velocity of the vehicle. This Doppler shifted signal when reflected off the vehicle is shifted again (due to the motion of the vehicle) back to the radar. A factor of 2 is introduced into the radar Doppler shift (see figure below). Note that the target velocity term (V) is positive for targets approaching the radar and negative for targets moving away from the radar (c is speed of light).

Figure C-1—**Stationary Radar Doppler Shift**

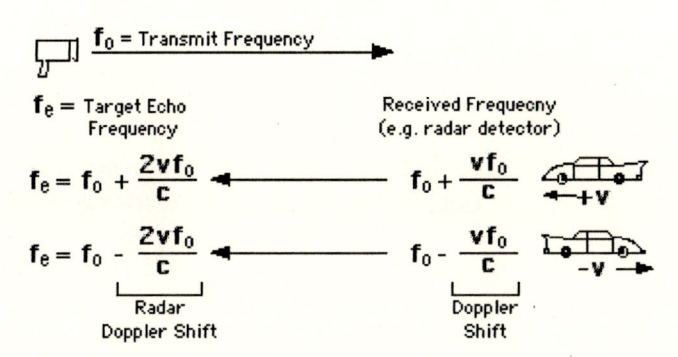

f_0 = Transmit Frequency

f_e = Target Echo Frequency

Received Frequency (e.g. radar detector)

$$f_e = f_0 + \frac{2vf_0}{c} \qquad f_0 + \frac{vf_0}{c}$$

$$f_e = f_0 - \frac{2vf_0}{c} \qquad f_0 - \frac{vf_0}{c}$$

Radar Doppler Shift · Doppler Shift

Moving mode target echoes are Doppler shifted by the target speed AND patrol car speed (closing or opening speeds between target and patrol car). See figure below. For patrol cars with radars pointed forward the patrol car velocity (Vp) is positive, target velocity (V) is positive for approaching targets and negative for targets traveling in the same direction. For patrol cars with radars pointed aft (rear) the patrol car velocity (Vp) is negative, target velocity (V) is positive for targets traveling in the same direction and negative for targets traveling in the opposite direction.

Donald S. Sawicki

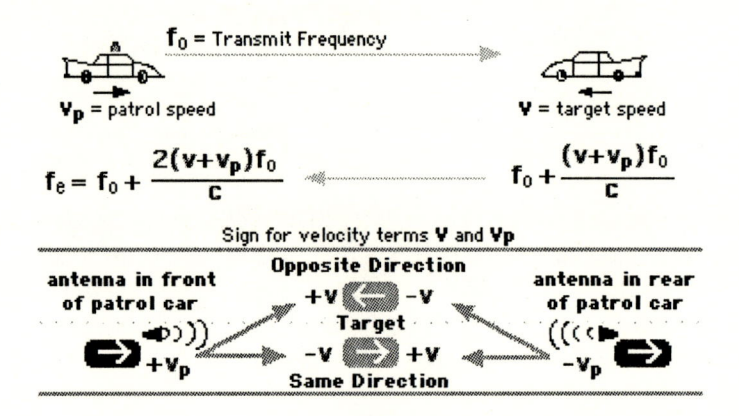

$$f_e = f_0 + \frac{2(v+v_p)f_0}{c} \qquad f_0 + \frac{(v+v_p)f_0}{c}$$

Figure C-2—**Moving Radar Doppler Shift**

The table below shows radar unit (Hz/mph and Hz/kmh) Doppler shift (either stationary radar or opening/closing speed for moving radar) per 1 mph (1.6 kmh) and 1 kmh (0.62 mph) for traffic radar (center) frequencies. Maximum measured speed to maintain a ± 1 mph (or kmh) error is based on transmit frequency and tolerance. The below figure plots radar Doppler shift versus target speed (or opening/closing speed for moving radar).

Radar Transmit Frequency		Radar Doppler Shift			Frequency Tolerance	Max Speed
Band	GHz	Hz/mph	Hz/kmh	Hz/knot	MHz	mph or kmh
S	2.455	7.29	4.53	8.39		
X	10.525	31.39	19.50	36.12	± 25	420
K	24.125	71.95	44.71	82.80	± 100	240
K	24.150	72.02	44.75	82.88	± 100	240
Ka	33.4	99.61	61.89	114.63	± 100	333
	to 36	107.36	66.71	123.55	± 100	359

Table C-1—**Unit Speed Doppler Shift and Max Speed**

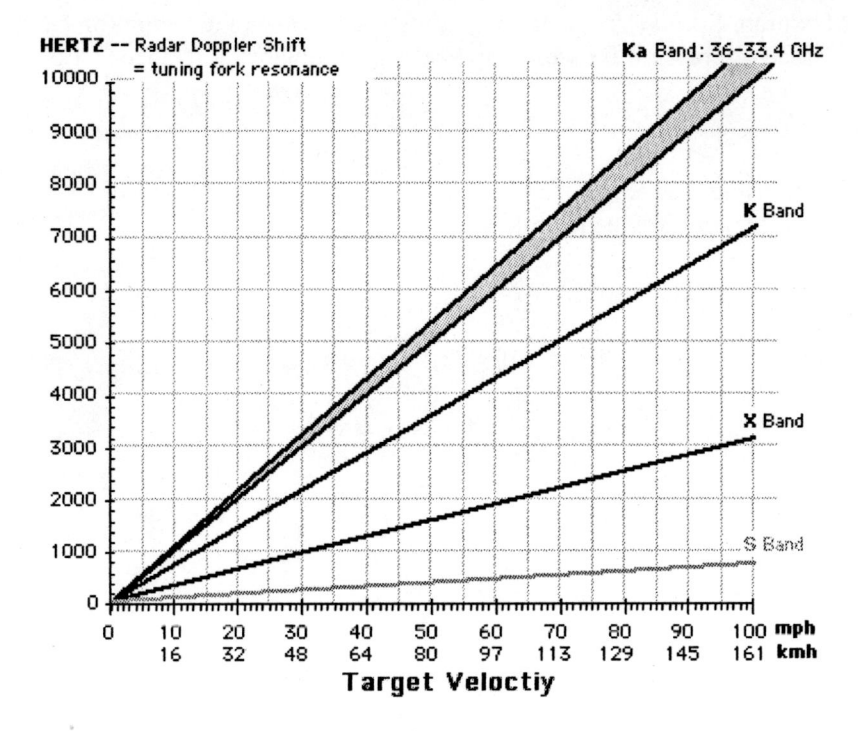

```
HERTZ -- Radar Doppler Shift
         = tuning fork resonance                    Ka Band: 36-33.4 GHz
10000

 9000

 8000
                                                              K Band
 7000

 6000

 5000

 4000
                                                              X Band
 3000

 2000

 1000                                                         S Band

    0
        0   10   20   30   40   50   60   70   80   90  100  mph
            16   32   48   64   80   97  113  129  145  161  kmh
                          Target Veloctiy
```

Figure C-3—**Radar Doppler Shift versus Speed**

Also see Tuning Forks (chapter 5.2—Testing the Radar) for radar Doppler shift range (bandwidth) ± 0.5 mph or kmh versus speed.

Also see The Doppler Principle (chapter 2.1).

Donald S. Sawicki

Maximum Speed (Vmax) to maintain a speed error of ± 1 mph (or 1 kmh) or less. Based on transmit frequency and transmit frequency tolerance (± ftol).

$$V_{max} < f_o / f_{tol}$$

f_o = transmit frequency
f_{tol} = ± transmit frequency tolerance

If ± x is acceptable speed error;

$$2v(f_o \pm f_{tol})/c < 2(v \pm x)f_o/c$$

$$v(f_o \pm f_{tol}) < (v \pm x)f_o$$

$$\mathbf{v < xf_o / f_{tol}}$$

appendix D
Radar Range Equation

Microwave Traffic Radars

There exist hundreds of versions of the radar range equation. Below is one of the more basic forms for a single antenna system (same antenna for both transmit and receive). The target is assumed to be in the center of the antenna beam. The maximum radar detection range (Rmax) is;

$$R_{max} = [\; P_t\, G^2\, \lambda^2\, \sigma / (4\pi)^3\, P_{min}\;]^{0.25}$$

P_t = transmit power (power dimensions)
G = antenna gain (ratio)
λ = transmit wavelength (length dimensions)
σ = Target radar cross section (area dimensions)
P_{min} = minimum detectable signal (power dimensions)

Substituting $\lambda = c / f_o$

$$R_{max} = [\; P_t\, G^2\, c^2\, \sigma / f_o^2\, (4\pi)^3\, P_{min}\;]^{0.25}$$

c = speed of light
f_o = transmit frequency (Hertz)

The variables in the above equation are constant and radar dependent except target RCS. Transmit power will be on the order of 1 mW (0 dBm) and antenna gain around 100 (20 dB) for an effective radiated power (ERP) of 100 mW (20 dBm). Minimum detectable signals are on the order of picowatts; RCS for an automobile might be on the order of 100 square meters. The accuracy of the radar range equation is only as good as the input data.

$$P_{min} = k\, T\, B\, (NF)\, (S/N)_{min}$$

$k = 1.38 \times 10^{-23}$ (watt*sec/°Kelvin)—Boltzmann's Constant
T = Temperature (°Kelvin)
B = receiver Bandwidth (Hertz)
NF = Noise Figure (ratio)
S/N_{min} = minimum Signal-to-Noise (ratio)

Minimum detectable signal depends on receiver bandwidth, noise figure, temperature, and required signal-to-noise ratio. A narrow bandwidth receiver will be more sensitive than a wider bandwidth receiver. Noise figure is a measure of how much noise a device (the receiver) contributes to a signal: the smaller the noise figure, the less noise the device contributes, and the better the sensitivity. Increasing temperature will affect receiver sensitivity by increasing input noise: the greater the temperature, the more noise, and the less receiver sensitivity. The available input thermal noise power (background noise) is proportional to the product kTB where k is Boltzmann's constant, T is temperature (degrees Kelvin) and B is receiver noise bandwidth (approximately receiver bandwidth). The product of kTB is noise power density for a given bandwidth.

The radar range equation above can be written for power received (Prec) as a function of target range for a given transmit power, wavelength, antenna gain, and target RCS.

$$\text{Prec} = Pt\, G^2 \lambda^2 \sigma / (4\pi)^3\, R^4$$
$$= Pt\, G^2 c^2 \sigma / fo^2\, (4\pi)^3\, R^4$$
$$= Pt\, G^2 (\lambda/4\pi R)^2 (\sigma/4\pi R^2)$$

$(\lambda/4\pi R)^2$—Range loss between transmit antenna and target
$(\sigma/4\pi R^2)$—loss due to target reflection and range

Note that radar has a range loss inversely proportional to range to the 4th power. Radio communications range losses are inversely proportional to range squared (one-way path). Signal power received (by a radar detector), where Gdet is detector antenna gain, can be expressed as shown below. By substituting radar detector minimum signal for power received, detector maximum range can be estimated if radar power and antenna gain are known (ERP—effective radiated power).

$$\text{Pdet} = Pt\, G\, \text{Gdet}\, \lambda^2 / (4\pi R)^2$$
$$= Pt\, G\, \text{Gdet}\, (\lambda/4\pi R)^2$$
$$= Pt\, G\, \text{Gdet}\, c^2 / fo^2\, (4\pi R)^2$$

$$R = \{ (Pt\, G\, \text{Gdet}\, \lambda^2) / [(4\pi)^2\, \text{Pdet}] \}^{1/2}$$
$$= \{ (Pt\, G\, \text{Gdet}\, c^2) / [(4\pi\, fo)^2\, \text{Pdet}] \}^{1/2}$$

Traffic Radar Handbook
A Comprehensive Guide to Speed Measuring Systems

appendix E
Radar Technical Details

Outline
- Introduction
- Antenna
   —Gain
   —Far-field
- Diplexer
- Transmitter
- Phase-Lock Loop (PLL)
- Receiver
- Processing

Basic hardware is similar in most radar models. Differences do exist; some models are for stationary use only, and some are for both stationary and moving mode. Some models are single units, and some are two or more pieces (boxes) and/or have multiple antennas. Many units measure only approaching targets; others measure both approaching and receding targets. Some moving mode radars have the ability to measure same-lane (direction) traffic (front and/or rear of patrol car). The below figure represents a typical system for traffic radar. For multi-piece radars the antenna is usually separate from the rest of the electronics.

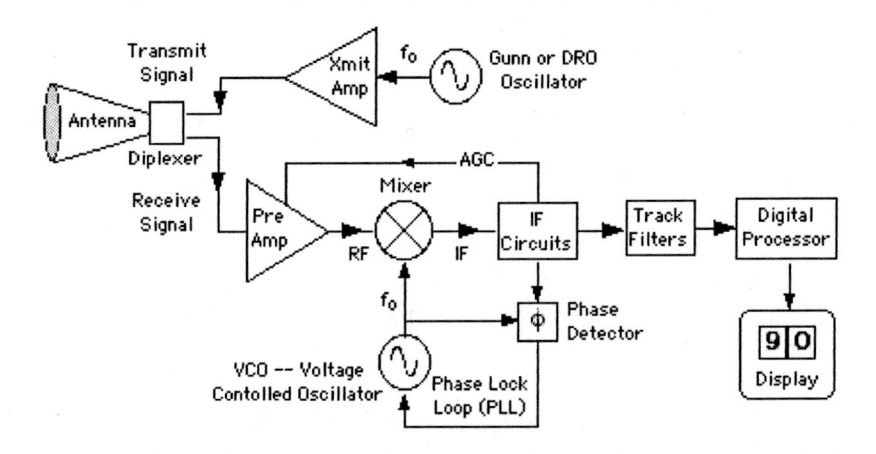

Figure E-1—**Radar Block Diagram**

Donald S. Sawicki

Antenna

GAIN

Antenna gain (G) can be expressed in terms of wavelength (λ) and effective antenna area (A_e). Effective antenna area is physical area (A) times efficiency (η).

$A_e = \eta A$

$$G = A_e 4\pi/\lambda^2 = \eta A\, 4\pi/\lambda^2$$
$$G_{dB} = 12 + 10 \log(A_e/\lambda^2) = 12 + 10 \log(\eta A/\lambda^2)$$

G = antenna Gain (ratio)

G_{dB} = antenna Gain in dB (decibels)

A_e = antenna Effective area

λ = Wavelength (distance dimensions)

η = aperture efficiency (ratio)

A = aperture physical area

Efficiency for a **pyramidal horn** antenna is typically **51.1%** (optimum), but can be as great as 80 percent (a very long flare length). Gain for an optimum pyramidal horn can be estimated from aperture horizontal and vertical dimensions (a and b). Note the product of a and b (a x b) is area.

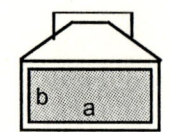

$$G = 4\pi\eta a\, b/\lambda^2 = 12.57\, \eta\, (a/\lambda)(b/\lambda)$$
$$G_{dB} = 11 + 10 \log[\eta\,(a\,b/\lambda^2)]$$

$$G = 6.28\, a\, b/\lambda^2$$
$$G_{dB} = 8.08 + 10 \log[\eta\,(a\,b/\lambda^2)]$$
$$\eta = 51.1\ \%\ (\text{optimum})$$

Optimum gain for a **conical horn** can be estimated from aperture circumference (C), wavelength, and efficiency (typically **52.2%,**).

$$G = \eta\,(C/\lambda)^2$$
$$G_{dB} = 10 \log(\eta) + 20 \log(C/\lambda)$$

$$G = 0.52 \, (C/\lambda)^2$$
$$G_{dB} = -2.82 + 20 \log (C/\lambda)$$
$$\eta = 52.2 \% \text{ (optimum)}$$

Gain for a **parabolic reflector** can be estimated from the plane circular area (A) the reflector covers. Note, η is typically 50% to 65%.

$$G = 4\pi\eta A / \lambda^2$$
$$G_{dB} = 11 + 10 \log (\eta) + 10 \log (A/\lambda^2)$$

FAR-FIELD

Antenna gain and beamwidth are antenna (pattern) parameters that apply only at long ranges. Long ranges are defined as any range that is greater than two times the diameter squared of the radar antenna diameter divided by the wavelength (see below figure).

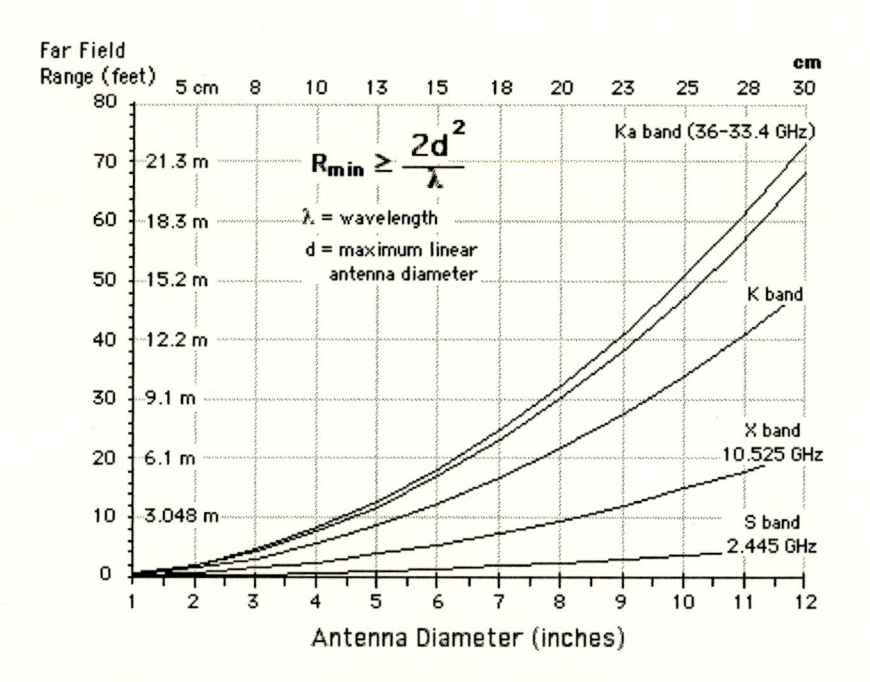

$$R_{min} \geq \frac{2d^2}{\lambda}$$

λ = wavelength
d = maximum linear antenna diameter

Figure E-2—**Antenna Far-field Region**

Donald S. Sawicki

The Fresnel region is defined as the range that is less than the far-field and greater than half the range of the minimum far field. The range less than half the distance of the far-field is the near-field region. In the Fresnel and near-field regions the beam is still forming (focusing).

Diplexer

An important component associated with a shared transmit/receive antenna is the diplexer. A diplexer is a passive device that routes the transmit signal to the antenna and not the receiver, and routes signals from the antenna to the receiver and not the transmitter. The diplexer is not perfect and leaks signals to other ports, but leakage is greatly attenuated if the device is designed properly. However, some transmitter leakage is required for use by the receiver in order to fine-tune (with a PLL—phase-locked loop) the receiver VCO (Voltage Controlled Oscillator) to compensate for any transmit oscillator frequency drift.

Transmitter

A Gunn (diode) oscillator or a DRO (dielectric resonant oscillator) can be used to generate the radar microwave transmit signal. A Gunn oscillator is a solid-state device and a DRO is a ceramic device. An amplifier may be required to increase signal power before transmission with either device. DROs are less stable (in frequency) and seldom used as a transmit signal source.

Phase-Lock Loop (PLL)

Traffic radar is designed to detect transmitter leakage to track any frequency drift of the Gunn or DRO oscillator. If the transmit frequency drifts, the receiver electronically tracks this drift with a phase-lock loop (PLL) and tunes the receiver to the proper frequency by adjusting the VCO (voltage controlled oscillator) frequency.

Receiver

The AGC (automatic gain control) circuits adjust receiver gain (sensitivity) to prevent receiver saturation from strong signals, such as strong ground echoes or large close targets. However when the AGC lowers the receiver gain, targets at longer ranges and/or smaller targets may not be detected.

Target return signals are enhanced by a low noise amplifier (LNA). This amplifier differs from the transmit amplifier in that the transmit amplifier is intended to boost a signal of moderate power as much as possible (noise gets amplified too) while the LNA is intended to enhance a weak signal without amplifying or adding noise.

After the LNA processes the signal a mixer down converts the signal to an intermediate frequency (IF). The IF is the difference between the target return frequency and the local oscillator frequency. Some moving-mode radars adjust the local oscillator frequency proportionally to the patrol car velocity in order to subtract the radar's velocity (patrol car speed) from target returns. This allows the same IF circuits and track filters to process both stationary and moving modes.

Processing

The IF circuits process signal returns for use by the track filters. The track filters separate the returns to detect discrete signals for power and frequency. Any signals detected are sampled and tracked by a digital processor. If multiple targets are present, the processor determines which ONE (or two) to display. Microwave radars update the display approximately one to four times per second depending on model. Radar integration time (sample time period) is equal to or greater than (whole number multiple) of display update time.

Donald S. Sawicki

appendix F
Ladar Technical Details

Outline
- Minimum Range
- Target Velocity
- Block Diagram

Ladars are really laser range finders adapted to calculate target velocity by measuring range difference over a period of time. Ladars are pulse modulated—transmit laser light pulses at a fixed rate. The ladar transmits a short light pulse and listens for an echo return to measure pulse round trip travel time (to calculate range). At least two range measurements (2 pulses) are required to establish target velocity—in practice tens to hundreds of pulses are processed to determine target speed.

Minimum Range
During pulse transmit time the receiver must be blanked (shut down) to prevent excess light energy from disrupting the receiver. Blanking imposes a minimum range on the ladar that is a function of pulse width (see figure below). The pulse will have a turn on time (rise time) and turn off time (fall time) that also requires the receiver be blanked. To receive an entire pulse (as opposed to a fraction), target range must be at least half a pulse width (moves minimum range to 3/2 pulse width). Close range reflections could have sufficient energy to saturate or damage the receiver and may require additional blank time.

Donald S. Sawicki

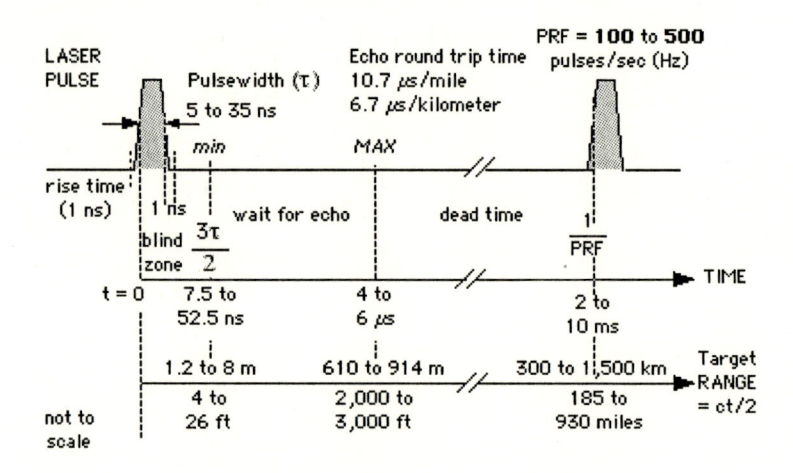

Figure F-1—Ladar Pulse Parameters

PRF - Pulse Repetition Frequency (pulses per second or Hz)
1 ns = 1 nanosecond = 0.000 000 001 seconds
1 μs = 1 microsecond = 0.000 001 seconds
1 ms = 1 millisecond = 0.001 seconds

Target Velocity

Because the transmit pulse travels twice the distance of target range, a factor of 2 is introduced in equations relating time and target range. The time it takes a pulse to travel from the ladar to the target (where c is speed of light) is (RANGE x c). The time it takes a pulse to travel from the ladar to the target and back is (2 x RANGE x c). Ladar target range is (where t = pulse round trip travel time):

Range = c t / 2, t = 2 (Range) / c
t = pulse round trip travel time, c = speed of light

One microsecond (0.000001 sec) equals a radar range of approximately 500 feet (pulse travels 1,000 feet); one nanosecond (0.000000001) equals a radar range of approximately 0.5 feet (pulse travels 1 foot).

The ladar collects all data points (echoes) during a sample time before calculating target velocity (see figure below). To qualify as a valid (single) data point, criteria might include echo power be sufficient, and the pulse width and/or wavelength of the echo be within expected bounds. An entire data set (1 sample time) may be ignored (discarded) if too many returns are missing, or are too far

Traffic Radar Handbook
A Comprehensive Guide to Speed Measuring Systems

off (time-range) from other data in the set, or do not meet other defined limits. The entire data set must also meet predetermined variance and standard deviation constraints. A valid data set is usually processed with a least squared average (or other smoothing algorithm) to smooth out minor range errors before calculating target velocity.

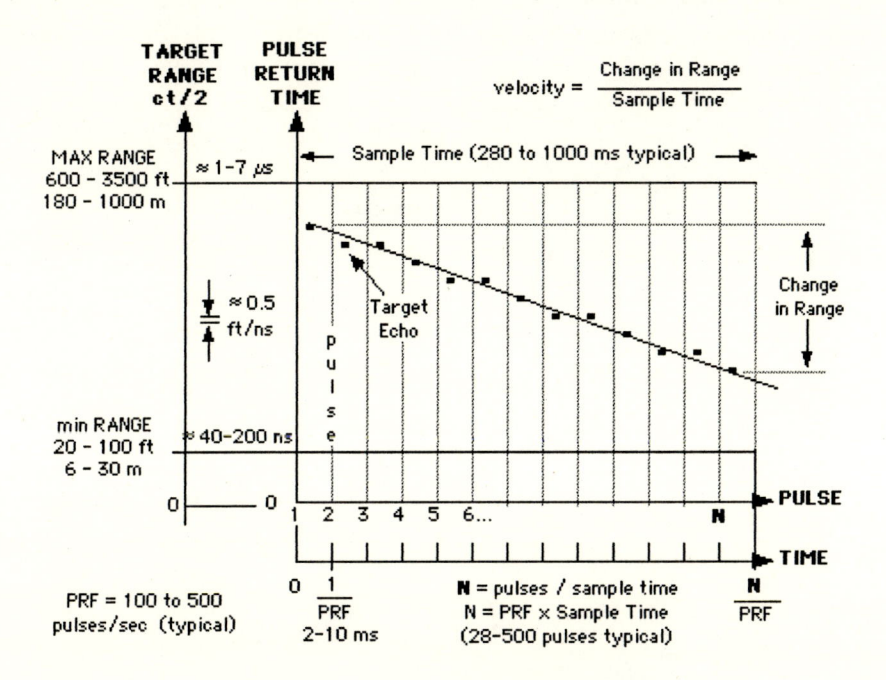

Figure F-2—**Ladar Timing Parameters**

If a data set is considered valid, the ladar calculates target velocity from change in range divided by sample time. If target range is decreasing with time, the target is approaching the ladar. If target range is increasing with time, the target is going away from the ladar. Under ideal conditions the ladar could display a speed reading after only one sample period. In practice a ladar may take as long as 1 second (about 3 sample periods) to display target speed. Bad conditions (fog, rain, carbon dioxide, obstructions) could force the ladar to take as long as several seconds or miss a target entirely.

Block Diagram
The figure below is a general block diagram for a ladar system. There will be variations from this diagram in practice, but it does illustrate the basic form and principles. The laser will generate light pulses and route them to the

aperture. Note some ladars use a single aperture for transmit/receive, some use separate apertures. Return signals route directly from the aperture to a light detector. The detector, which may or may not include a light amplifier, converts light detections to electrical energy and routes the signals to a video amplifier. The video amplifier adjusts signal gain using STC (sensitivity time control) curves (longer range signals require more gain) for the threshold detection circuits. Video signals that exceed the detection threshold get processed for leading edge detection or range gates (range tracking). If the data set passes the preprogrammed test criteria, target data can then be calculated and displayed.

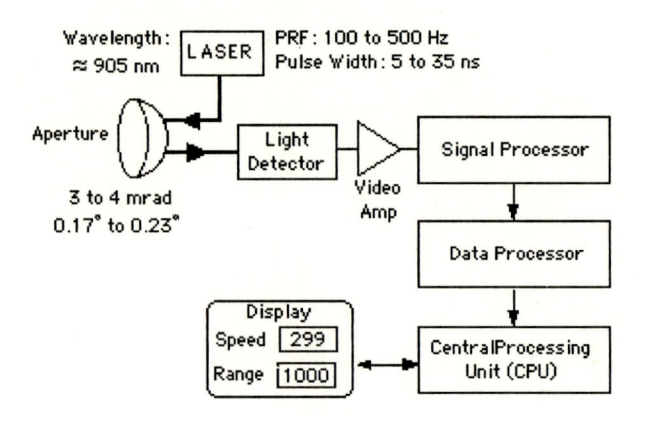

Figure F-3—**Ladar Block Diagram**

Ladars use a programmable CPU (Central Processing Unit). A programmable processor allows the flexibility to program different pulse widths, PRF's, sample time, and algorithms (for computing velocity). Exact parameters may vary somewhat from manufacturer to manufacturer, or model to model, or even software version to software version. Pulse transmit rates (PRFs) could be as high as 100,000 Hz in theory (0.9 miles), but in practice will probably be on the order of several hundred Hertz (100 to 500 pulses per second). Sample times will be on the order of 0.3 to 1 seconds. Pulse widths have been reported to be as low as 5 ns and as high as 35 ns.

Traffic Radar Handbook
A Comprehensive Guide to Speed Measuring Systems

Constants / Conversions / Abbreviations

OUTLINE
- Decimal Multiplier Prefixes
- Constants
- Decibels (dB)
- Speed of Light
- Speed Conversions
- Length Conversions
- Temperature

Decimal Multiplier Prefixes

MULTIPLIER	POWER	PREFIX	SYMBOL	
1,000,000,000,000	10^{12}	tera	T	trillion
1,000,000,000	10^{9}	giga	G	billion
1,000,000	10^{6}	mega	M	million
1,000	10^{3}	kilo	k	thousand
100	10^{2}	hecto	h	
10	10^{1}	deka	da	
0.1	10^{-1}	deci	d	
0.01	10^{-2}	centi	c	
0.001	10^{-3}	milli	m	
0.000 001	10^{-6}	micro	µ	
0.000 000 001	10^{-9}	nano	n	
0.000 000 000 001	10^{-12}	pico	p	
0.000 000 000 000 001	10^{-15}	femto	f	
0.000 000 000 000 000 001	10^{-18}	atto	a	

In some (European) countries 10^{18} is a trillion, and 10^{12} is a billion.

Constants

pi = π = 3.141 592 653 589... $180° = \pi$ (radians)	1 rad = $180°/\pi \approx 57.3°$ $1° = (180°/\pi) \approx 0.01745$ rad
permeability of free space $\mu_o = 4\pi \times 10^{-7}$ (Henrys/meter)	**permittivity of free space** $\varepsilon_o \approx 8.854\ 185 \times 10^{-12}$ (Farads/meter)
Boltzmann's Constant k = 1.38×10^{-23} (watt*sec/T°k)	$\approx$ Approximately

Decibels (dB)
W = watts, mW = milliwatts

$$\text{ratio} = 10^{dB/10} \qquad dB = 10\ \log_{10}(\text{ratio})$$

$$mW = 10^{dBm/10} \qquad dBm = 10\ \log_{10}(mW)$$

$$W = 10^{dBW/10} \qquad dBW = 10\ \log_{10}(W)$$

$$3\ dB = 1.9952 \approx 2$$
$$-3\ dB = 0.5012 \approx 1/2$$

RADIO Range
+6 dB difference equates to an increase in range of 2 times,
-6 dB difference equates to a decrease in range of 1/2.

RADAR Range
+12 dB difference equates to an increase in range of 2 times,
-12 dB difference equates to a decrease in range of 1/2.

Traffic Radar Handbook
A Comprehensive Guide to Speed Measuring Systems

Speed of Light (c) in a vacuum

299,792	kilometers / second	(km / s)
1,079,252,842	kilometers / hour	(km / hr or kmh)
186,282	miles / second	(mi / s)
670,616,625	miles / hour	(mi / hr or mph)
983,571,050	feet / second	(ft / s)
983.571	feet / microsecond	(ft / μs)
0.983	feet / nanosecond	(ft / ns)
161,875	nautical miles / second	(NM / s)
582,749,914	nautical miles / hour	(NM / hr or knots)

Approximate Speed of Sound at sea level
1,230 km / hr
765 mph
665 knots

Speed Conversions

1 foot / second =	0.304 8 (exactly)	m / s
	0.592 483 801	knots
	0.681 818 182	mph
	1.097 28 (exactly)	km/hr (kmh)
1 kilometer / hour =	0.277 777 778	m / s
	0.539 956 803	knots
	0.621 371 192	mph
	0.911 344 415	ft / s
1 mile / hour or mph =	0.447 04 (exactly)	m / s
	0.868 976 242	knots
	1.466 666 667	ft / s
	1.609 344 (exactly)	km/hr (kmh)
1 NM / hr or knot =	0.514 444 444	m / s
	1.150 779 448	mph
	1.687 809 857	ft / s
	1.852 (exactly)	km/hr (kmh)

Donald S. Sawicki

Length Conversions

1 inch =	0.025 4 (exactly)	meters
1 meter =	1.093 613 298	yards
	3.280 839 895	feet
1 kilometer =	0.539 956 803	nautical miles
	0.621 371 192	miles
	1,093.613 298	yards
	3,280.839 895	feet
1 mile (statute) =	0.868 976 242	nautical miles
	1.609 344	kilometers
	1,760 (exactly)	yards
	5,280 (exactly)	feet
1 nautical mile =	1.150 779 448	miles
	1.852 (exactly)	kilometers
	2,025.371 829	yards
	6,076.115 486	feet

Temperature

Tk = temperature in degrees Kelvin	TF = temperature in degrees Fahrenheit	Tc = temperature in degrees Celsius
Tk = Tc + 273.15° = (5/9)(TF + 459.67°)	Tf = (9/5)Tc + 32° = (5/9)(Tk) − 459.67°	Tc = (5/9)(TF − 32°) = Tk − 273.15°

Additional Reading / Sources

Outline
    Data References
    Periodicals
    Text Books
    Traffic Radar Books
    U.S. Government Documents

DATA REFERENCES

1. American National Standard Safety Levels with Respect to Human Exposure to Radio Frequency Electromagnetic Fields (300 kHz - 100 GHz), ANSI Standard C95.1- 1982.

2. Compliance Engineering, 1999 Annual Reference Guide, Volume XVI, No.4, Compliance Engineering, 1999.

3. Compliance Engineering, 1997 Annual Reference Guide, Volume XIV, No. 3, Compliance Engineering, 1997.

4. CRC Standard Mathematical Tables, 20th Edition, Samuel M. Selby, Ph.D. Sc.D., CRC Press, 1972.

5. International System of Units (SI), Physical Constants and Conversion factors 2nd revision, E. A. Mechtly - University of Illinois, NASA SP-7012, 1973

6. Occupational Exposure of Police Officers to Microwave Radiation from Traffic Radar Devices W. Gregory Lotz, Robert A. Rinsky, Richard D. Edwards, National Technical Information Service (NTIS) publication number PB95-261350, June, 1995.

PERIODICALS

1. Are mobile phones safe?, Kenneth R. Foster (U of PA) and John E. Moulder (Med College of WI), p 23-28, Institute of Electric and Electronic Engineers (IEEE) Spectrum magazine, Aug 2000.

2. Ask Bob (police radar frequencies), page 98, Monitoring Times, March 1991.

3. Big Brother Is Driving, Time, November 23, 1953.

4. The Cellular Phone Scare, Mark Fischetti, Institute of Electric and Electronic Engineers (IEEE) Spectrum magazine, June 1993

5. Collision avoidance keys on wireless technology to make highways intelligent, Roger Lesser, RF Design, p 72-76, October 1998

6. Critics Blast Gov't Agency For Refusing To Take Aim At Radar-Gun Makers, Electronic Engineering TIMES, December 21, 1981.

7. Electromagnetic fields: the jury's still out, Institute of Electric and Electronic Engineers (IEEE) Spectrum magazine, August 1990.

8. EPA examines electromagnetics, Richard Doherty, Electronic Engineering TIMES, September 17, 1990.

9. FCC Slams Jammers, RADAR Reporter Newsletter, winter 1998.

10. Frequency sweeps improve radar, Electronic Engineering Times, p. 76, July 12, 1999.

11. The friendly fields of RF, Robert E. Shupe and Neb B. Hornback, Institute of Electric and Electronic Engineers (IEEE) Spectrum magazine, June 1985.

12. GADGETS, Burble & Squeak, Time, January 19, 1962.

13. Greyhound Gives Up On VORAD, RADAR Reporter Newsletter, September 1995.

14. Helping speeders beat the radar rap, Institute of Electric and Electronic Engineers (IEEE) Spectrum magazine, August 1978.

15. Improving on police radar, P. David Fisher - Michigan State University, Institute of Electric and Electronic Engineers (IEEE) Spectrum magazine, July 1992.

16. Interceptor VG-2, Radar Detector Detector, RADAR Reporter Newsletter, September 1990.

17. Invisible Villains - the Product of Electromagnetic Profusion, Insight, July 4, 1988.

18. Is Highway Radar Foolproof?, Charles Remsberg, Popular Mechanics, June 1962.

19. It's the Photo and Laser Cops, Don Sherman, Popular Science, September 1990.

20. Memory in megabytes and/or mebibytes, Institute of Electrical and Electronic Engineers (IEEE) Spectrum magazine, August 1999.

21. Millimeter-wave energy to be used in a weapon, Peter Clarke, Electronic Engineering TIMES, p 26, June 11, 2001.

22. N.Y. limits detector use, Richard Doherty, Electronic Engineering TIMES, August 20, 1990.

23. Nearly circular beams spark red lasers, Yoshiko Hara, Electronic Engineering TIMES, July 9, 2001.

24. Police Traffic - Radar Units Found Deceptive At Confab, Richard Doherty, Electronic Engineering TIMES, November 5/12, 1979.

25. Radar Jammers: Bogus and Bonafide, RADAR Reporter Newsletter, 1993 NOV.

26. Radar Really Catches Speeders, U.S. News & World Report, August 06, 1954.

27. Radio Waves and the Human Body: Two Interesting Views, by Bob Grove—Publisher, p 104, Monitoring Times, April 1999.

28. RF radiation: biological effects, Institute of Electrical and Electronic Engineers (IEEE) Spectrum magazine, December 1980.

29. Shortcomings of radar speed measurement, Richard Doherty, Institute of Electrical and Electronic Engineers (IEEE) Spectrum magazine, December 1980.

30. Slow Down - Radar Ahead!, Don Wharton, Readers Digest, January 1966.

31. System Considerations for the Design of Radar Braking Sensors
Richard A. Chandler and Lockett E. Wood, IEEE Transactions on Vehicular Technology, vol. VT-26, No. 2, pp 151-160, May 1977.

32. They have Lasers, Patrick Bedard, Car and Driver, April 1992.

TEXT BOOKS

1. Antenna Engineering Handbook, 1st Edition, Henry Jasik, McGraw-Hill, 1961.

2. Digital Signal Processing, Alan V. Oppenheim and Ronald W. Schafer, Prentice-Hall, 1975.

3. Fundamentals of Waves, Optics and Modern Physics, 2nd Edition, Hugh D. Young, McGraw-Hill, 1976.

4. Introduction to AIRBORNE RADAR, George W. Stimson, Hughes Aircraft Co. - Radar Systems Group, CA, 1983.

5. Introduction to Radar Systems, 2nd Edition, Merrill I. Skolnik, McGraw-Hill, 1980.

6. Radar Handbook, Merrill I. Skolnik, McGraw-Hill, 1970.

TRAFFIC RADAR BOOKS

1. Beating The Radar Rap, 2nd Edition, Dale T. Smith & John Tomerlin, Bonus Books Inc., Chicago, 1990.

2. Case Dismissed II, Radio Association Defending Airwave Rights, Inc. (R.A.D.A.R.), 1997.

3. Speed Monitoring Technology Handbook, Radio Association Defending Airwave Rights, Inc. (R.A.D.A.R.), 1996.

4. The Traffic Radar Handbook—A Comprehensive Guide to Speed Measuring Systems 1st Edition, Donald S. Sawicki, Grove Enterprises, Inc. 1993 OCT, ISBN 0-944543-08-1

U.S. GOVERNMENT DOCUMENTS

1. FCC Public Notice—FCC Regulates Radar Transmitters, but not Radar Detectors Delegated Authority DA 96-2040, 1996 DEC 09.

2. FCC Rules and Regulations, Parts 15, 20, and 90.

3. Performance Standards for Speed Measuring Devices, United States Department of Transportation and National Highway Traffic Safety Administration, Federal Register Volume 46, Number 5, January 8, 1981.

4. Police Traffic Radar Issue Paper, US Department of Transportation & National Highway Traffic Safety Administration, DOT HS-805 254, February 1980.

Donald S. Sawicki

About the Author
Donald S. Sawicki

PUBLICATIONS
- **Traffic Radar Handbook Web** Site [CopRadar.com]
  A Comprehensive Guide to Speed Measuring Systems, 1998 May
  TXu 824-435, 11-18-97, 096052004
- **The Traffic Radar Handbook** (out of print)
  A Comprehensive Guide to Speed Measuring Systems
  Published by Grove Enterprises, 1993 Oct, ISBN 0-944543-08-1
- **F-15 Radar / Jammer Interoperability**
  Published by the Environment Research Institute of Michigan
  (ERIM), 1990 May

PRESENTATIONS
- Speaker at the 35th Annual **Tri-Service Radar Symposium**, 1989 Jun
  U.S. Military Academy, West Point, NY

EXPERIENCE
- **Lead Engineer**—Advanced Design, McDonnell Aircraft, 1984-91
  Electronic Warfare Systems Analysts for Advanced Design and F-15
  A/E Eagle, Systems Engineer for APG-65 radar (F/A-18 A/D Hornet)
- **Senior Engineer**—Coherent Radar Systems, Emerson Electric 1980-84
  Systems Engineer for APG-69 radar design and F-5F (Tiger) integration,
  APQ-157/159 radar enhancements and system integration.
- **Member Technical Staff**—Radar Systems, Hughes Aircraft, 1978-80
  Systems / Flight Test Engineer for GBU-15 (guide bomb unit).

EDUCATION
- Bachelor of Science Electrical & Electronic Engineering
  University of Illinois—Champaign / Urbana, Illinois (1978)

ORGANIZATIONS
- Institute of Electrical and Electronic Engineers (1977 - present)

COURT RECOGNITION
Courts that have recognized the author as a radar expert with expertise in traffic radar.

- State of Illinois, county of Madison, 3rd Judicial Circuit
- State of Missouri, county of St. Louis, 21st Judicial Circuit—Division 31
- State of Missouri, county of St. Louis, 21st Judicial Circuit—Division 33

ILLINOIS DRIVER SAFETY CITATIONs
IN RECOGNITION OF A PERSONAL CONTRIBUTION TO SAFE DRIVING…
without a Moving Traffic Violation (that could be proven) or Accident.

- IL Safety Citations: 1981-84, 1984-87, 1987-91, 1991-95, 1995-99. TOTAL: 18 consecutive years.

Printed in the United States
26519LVS00003B/138